Emerging Technologies in Architecture

Matthew N. O. Sadiku

Emerging Technologies in Architecture

Springer

Matthew N. O. Sadiku
Prairie View A&M University
Prairie View, TX, USA

ISBN 978-3-031-99508-8 ISBN 978-3-031-99509-5 (eBook)
https://doi.org/10.1007/978-3-031-99509-5

This Springer imprint is published by the registered company Springer Nature Switzerland AG
The registered company address is: Gewerbestrasse 11, 6330 Cham, Switzerland

Dedicated to my friends:
Joshua A. Efuwape
Judith T. Efuwape

Preface

Architecture is the art and science of designing buildings and other physical structures. The purpose of architecture is to serve the people and make day-to-day life more manageable. Technology has the capacity to transform the work of architects, and as a community they should approach its offerings with openness, persistence, optimism, and skepticism. In the digital world, builders have seen technology as a means to optimize practices and identify, build, and manage their projects. Without new technologies, designers would not be able to accommodate the more complicated demands of the industry. Technology is altering not just the architecture industry but the role of an architect. Modern architecture is characterized by freedom of ideas mainly due to the development and implementation of new technologies for architects.

There is no denying the technological transformation that architecture is currently undergoing. From design software to electronic utilities and applications, architecture is growing into one of the most technically savvy professions in the world. The architecture, engineering, and construction (AEC) industry is at the forefront of a technological revolution that will redefine how we design, build, and experience the built environment. In the ever-evolving realm of architecture, the integration of cutting-edge technologies has revolutionized the way architects and designers conceptualize, plan, and communicate their ideas. As technology continues to rapidly advance, architects need to stay ahead of the curve and incorporate cutting-edge tools into their design processes.

This book examines emerging technologies and their impact on architecture. It is organized into twelve chapters that summarize emerging technologies and their applications in architecture.

Chapter 1 "Introduction" The primary objective of this chapter is to comprehensively investigate the challenges and potentials of the adoption of emerging technologies in modern architecture. The chapter considers emerging technologies such as building information modelling (BIM), artificial intelligence (AI), cloud computing, big data, blockchain, and 3D printing. Emerging technologies increasingly influence architects' work, acting as catalysts for creativity, productivity, and communication.

These new architecture technologies are transforming how buildings are made and help design the buildings of tomorrow, today.

Chapter 2 "Building Information Modeling" This chapter is intended to provide the use of building information modeling (BIM) in the architecture industry. BIM is a way of representing a built asset as a digital model. It is the foundation of digital transformation in the architecture, engineering, and construction (AEC) industry. While BIM is a game-changer for the entire construction industry, it has particular benefits for architects. It allows architects, engineers, and other stakeholders to explore every aspect of a building–foundations, surroundings, past configurations, and even possible futures. BIM has emerged as a transformative force, redefining the way buildings are designed, constructed, and managed.

Chapter 3 "Artificial Intelligence in Architecture" This chapter examines the role of artificial intelligence (AI) in the architecture industry. AI encompasses a wide range of technologies that mimic human cognitive functions like problem-solving and learning from experiences. It refers to computer systems that can perform tasks traditionally done by humans. It is an emerging cornerstone in the field of architectural design. The impact of AI on architecture has been remarkable, leading to an improved workflow efficiency, optimized resource utilization, and sustainable designs. The future of AI in architecture and design holds incredible promise, pushing the boundaries of architecture even further.

Chapter 4 "Generative Design in Architecture" This chapter examines the various uses of generative design in architecture. Generative design is a design exploration process that utilizes algorithms to generate and evaluate numerous design alternatives based on specified goals and constraints, allowing architects to discover innovative solutions that may not be immediately apparent through conventional design methods. It is a design method that uses algorithms and artificial intelligence (AI) to generate multiple design solutions based on specific requirements and constraints. Generative design is making a name in the world of contemporary architecture.

Chapter 5 "Robotics in Architecture" This chapter delves into the innovative ways robotics is being integrated into architecture. Robotics has always inspired a keen sense of curiosity and interest among individuals. It is the discipline of designing and constructing machines, called robots. A robot is an autonomous mechanical device that is designed to sense its environment, carry out computations to make decisions, and perform actions like humans in the real world. Robots are transforming the way architects approach their designs. They are already being used in virtually every step of the building design process, from initial site analysis to construction. Architects who embrace robotics now will be better equipped to design the most efficient buildings of the future.

Chapter 6 "Drones in Architecture" In this chapter, we focus on drone applications in the architecture industry. Unmanned aerial vehicles (UAVs), commonly called drones, are gaining in popularity among professionals working in the Architecture, Engineering, and Construction (AEC) industry. Drone technology is playing a key role in redefining how we maintain, build, and design properties. Architects mainly use drones for generating photographic views and developing their projects from particular angles. Having a drone at the disposal of an architectural practice allows architects to be proactive.

Chapter 7 "Internet of Things in Architecture" This chapter covers Internet of things (IoT) applications in the architecture industry. IoT is a term that describes a wide variety of devices that connect to computer networks such as the Internet. It is about turning physical objects into smart objects. IoT connects sensor-based objects to a shared infrastructure, which can be an extranet, the Internet, or a network that eliminates the need for a human-to-computer interaction to exchange data. It shows exciting promise for revolutionizing the functions of ACE in several ways. It is changing the architecture and design trades. The IoT technology has revolutionized how we design and build modern buildings.

Chapter 8 "Big Data in Architecture" This chapter examines the role of big data in architecture. Big data is a broad term for the collection of both structured and unstructured data. When properly analyzed, big data can reveal patterns, trends, and relationships related to human behavior and interaction in the built environment. Big data in architecture refers to the application of big data technologies and techniques within the field of architecture, where large volumes of data from building sensors, user interactions, and environmental factors are collected and analyzed to optimize building design, performance, and operations. Architects who can harness big data will undoubtedly lead the charge toward more responsive and responsible built environments.

Chapter 9 "Cloud Computing in Architecture" This chapter explores the ways by which cloud computing (CC) helps organizations in the architecture industry become more efficient, sustainable, and innovative. Cloud computing is the delivery of computing services, such as servers, storage, databases, networking, software, analytics, and intelligence, over the Internet. It also allows users to access additional storage, enhanced processing power, and hardware-heavy architecture software all remotely through the Internet. Architects are increasingly using cloud computing to access and manage design software and collaborate on projects in real-time with team members across different locations. Architects today are starting to use the cloud for more than just file sharing and storage.

Chapter 10 "Blockchain in Architecture" The aim of this chapter is to explore the use of blockchain technology in architecture. Blockchain is a technology that allows the recording of information in a way that it is difficult or practically impossible to alter, hack, or cheat. It is a decentralized system that uses a peer-to-peer network to create a digital ledger of transactions. Blockchain technology is still in its early stages of development and adoption in the design process in architecture. It has the potential to revolutionize the way architects design, communicate, and collaborate.

Chapter 11 " 3D Printing in Architecture" This chapter examines the use of 3D printing technology in the architecture industry. 3D printing, also known as additive manufacturing, is the process of creating physical items by layering materials based on the requirements of a digital model. It revolutionizes traditional architectural processes by layering materials to create three-dimensional structures from digital models. The architectural designing community has embraced 3D printing with open arms and harnessed its limitless potential. 3D printing is revolutionizing the field of architecture, and it is poised to change the way we build our cities and towns in the future.

Chapter 12 "Immersive Technologies in Architecture" In this chapter, we will explore how immersive technologies are reshaping our understanding of reality and virtual environments in architecture. Immersive technologies have the potential to impact the way we work, learn, and interact with one another. They allow architects to step inside their designs, engage clients in immersive experiences, collaborate more effectively, and enhance overall project efficiency. Architects and designers use these technologies to create virtual walkthroughs, 3D visualizations, and interactive models of buildings and spaces. The technologies enable modern architects to push the boundaries of creativity, offering endless possibilities for creating spaces that are not only visually striking but also highly functional and engaging.

This is a comprehensive introductory text on the issues, ideas, theories, and applications of emerging technologies in the architecture industry. It provides an overview of each technology in simple terms so that beginners can understand it. It is a must-read book for anyone who wants to learn about modern architecture, which have become vital in many areas of life.

Prairie View, USA Matthew N. O. Sadiku

Competing Interests The author has no competing interests to declare that are relevant to the content of this manuscript.

Contents

About the Author

Matthew N. O. Sadiku received his B. Sc. degree in 1978 from Ahmadu Bello University, Zaria, Nigeria and his M.Sc. and Ph.D. degrees from Tennessee Technological University, Cookeville, TN in 1982 and 1984 respectively. From 1984 to 1988, he was an assistant professor at Florida Atlantic University, Boca Raton, FL where he did graduate work in computer science. In total, he received seven college degrees. From 1988 to 2000, he was at the Temple University, Philadelphia, PA, where he became a full professor. From 2000 to 2002, he was with Lucent/Avaya, Holmdel, NJ as a system engineer and with Boeing Satellite Systems, Los Angeles, CA as a senior scientist. He is presently a Regents professor emeritus of electrical and computer engineering at Prairie View A&M University, Prairie View, TX.

He is the author of over 1,470 professional papers and over 160 books including *Elements of Electromagnetics* (Oxford University Press, 7th ed., 2018), *Fundamentals of Electric Circuits* (McGraw-Hill, 7th ed., 2020, with C. Alexander), *Computational Electromagnetics with MATLAB* (CRC Press, 4th ed., 2019), *Principles of Modern Communication Systems* (Cambridge University Press, 2017, with S. O. Agbo), and *Emerging Internet-based Technologies* (CRC Press, 2019). In addition to the engineering books, he has written Christian books including *Secrets of Successful Marriages, How to Discover God's Will for Your Life*, and commentaries on all the books of the New Testament Bible. Some of his books have been translated into ten languages: French, Korean, Chinese (and Chinese Long Form in Taiwan), Italian, Portuguese, Spanish, German, Dutch, Polish, and Russian.

He was the recipient of the 2000 McGraw-Hill/Jacob Millman Award for outstanding contributions in the field of electrical engineering. He was also the recipient of Regents Professor award for 2012–2013 by the Texas A&M University System. He is a registered professional engineer and a life fellow of the Institute of Electrical and Electronics Engineers (IEEE) "for contributions to computational electromagnetics and engineering education." He was the IEEE Region two Student Activities Committee Chairman. He was an associate editor for IEEE Transactions on Education. He is also a member of Association for Computing Machinery (ACM). His current research interests are in the areas of computational electromagnetic, computer science/networks, engineering education, and marriage counseling. His works can be found in his autobiography, *My Life and Work* (Author's Tranquility Press, 2024) or his website: www.matthew-sadiku.com. He currently resides with his wife Janet in Westlake, Florida. He can be reached via email at sadiku@ieee.org.

Chapter 1
Introduction

> *We shape our buildings; thereafter they shape us.—Winston Churchill.*

1.1 Introduction

Technology has become a powerful enabler. It has long reshaped the possibilities of architecture. When we talk about technology, we often think of robots, supercomputers, data centers or smartphones. But technology also refers to the invention of the first chipped stone tools. Technology is nothing more than the application of knowledge to achieve goals in a specific and reproducible way, for practical purposes. Technology is altering not just the architecture industry but the role of an architect. In the digital world, builders have seen technology as a means to optimize practices and identify, build, and manage their projects. Without new technologies, designers would not be able to accommodate the more complicated demands of the industry. Technology in architecture is depicted in Fig. 1.1 [1].

Digitalization in the architecture, engineering, and construction (AEC) sector is slow due to significant challenges in technology adoption. Although transforming the AEC sector with digital technologies (such as building information modelling (BIM), artificial intelligence (AI), cloud computing, big data, blockchain, virtual reality (VR)), new knowledge about technology implementations can help research and practice. The rapid evolution of digital technologies has brought forth an era of constant change and innovation.

The architecture, engineering and construction (AEC) industry is at the forefront of a technological revolution that will redefine how we design, build, and experience the built environment. As technology continues to rapidly advance, architects need to stay ahead of the curve and incorporate cutting-edge tools into their design processes. Technological integration has revolutionized the field of architecture, enabling architects to create more sustainable, efficient, and innovative buildings that meet the needs of modern society. It has transformed the field of architecture, enabling architects to create more sustainable, efficient, and innovative buildings. Figure 1.2 shows a

M. N. O. Sadiku, *Emerging Technologies in Architecture*,
https://doi.org/10.1007/978-3-031-99509-5_1

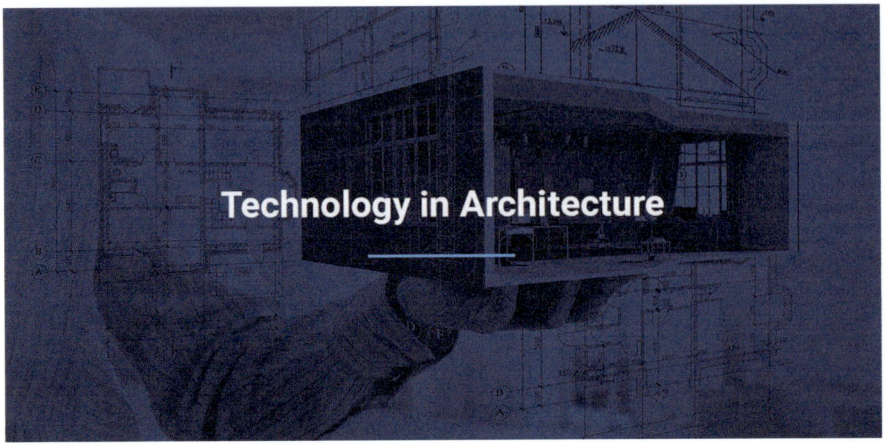

Fig. 1.1 Technology in architecture [1]

building under construction [2], while Fig. 1.3 displays some construction workers [3].

Modern architecture is characterized by freedom of ideas mainly due to the development and implementation of new technologies for architects. In our digital age, countless architectural innovation technologies are emerging and changing the way we design. Emerging technologies enable architects to implement ideas with greater precision and efficiency, reduce environmental impact, and create more sustainable and functional buildings. Emerging technologies are not just enhancing efficiency and precision; they are reshaping the very way we conceive, design, and construct

Fig. 1.2 A building under construction [2]

Fig. 1.3 Some construction workers [3]

the built environment. These exciting technologies drive the future of architecture and engineering [4].

The primary objective of this chapter is to comprehensively investigate the challenges and potentials of the adoption of emerging technologies in modern architecture. The chapter begins with explaining what emerging technologies are all about. It presents some specific emerging technologies in architecture. It highlights the benefits and challenges of these technologies. The last section concludes with comments.

1.2 What Are Emerging Technologies?

Technology may be regarded as a collection of systems designed to perform some function. It can help alleviate some of the challenges facing business today. Emerging technology is a term generally used to describe new technology. The term often refers to technologies currently developing or expected to be available within the next five to ten years. Any imminent, but not fully realized, technological innovations will have some impact on the status quo.

Emerging technologies are shaping our societies. They continue to affect the way we live, work, and interact with one another. Emerging technology (ET) lacks a consensus on what classifies them as "emergent." It is a relative term because one may see a technology as emerging and others may not see it the same way. It is a term that is often used to describe a new technology. A technology is still emerging if it is not yet a "must-have" [5]. An emerging technology is the one that holds the promise of creating a new economic engine and is trans-industrial. ET is used in different areas such as media, healthcare, business, science, education, or defense.

The characteristics of emerging technologies include the following [6]:

- *Novelty*: Emerging technologies are typically new or novel, meaning they have yet to be widely adopted or used. They often represent a significant departure from existing technologies or processes.
- *Potential for Disruption*: Emerging technologies have the potential to disrupt existing markets, industries, or ways of doing things. They may also displace existing businesses or industries.
- *Uncertainty*: Because emerging technologies are still in the early stages of development, there is often a high uncertainty surrounding their future potential and impact. It can be challenging to predict how they will evolve.
- *Rapid Change*: Emerging technologies often evolve rapidly, with new developments and innovations emerging frequently. It can make keeping up with the latest trends and advancements challenging.
- *Interdisciplinary*: Emerging technologies often involve multiple disciplines or fields of study, such as computer science, engineering, and biology. They may require collaboration across different fields and industries to develop their potential fully.

Emerging technologies are worth investigating. They are responsible for developing new products or devices. As emerging technologies continue to evolve, engineering is poised for a transformative future. Emerging technologies have driven innovation and progress in today's rapidly evolving digital landscape. The collective impact of emerging technologies such as artificial intelligence, machine learning, big data, and the Internet of things is undeniably transformative. Some emerging technologies are shown in Fig. 1.4 [7].

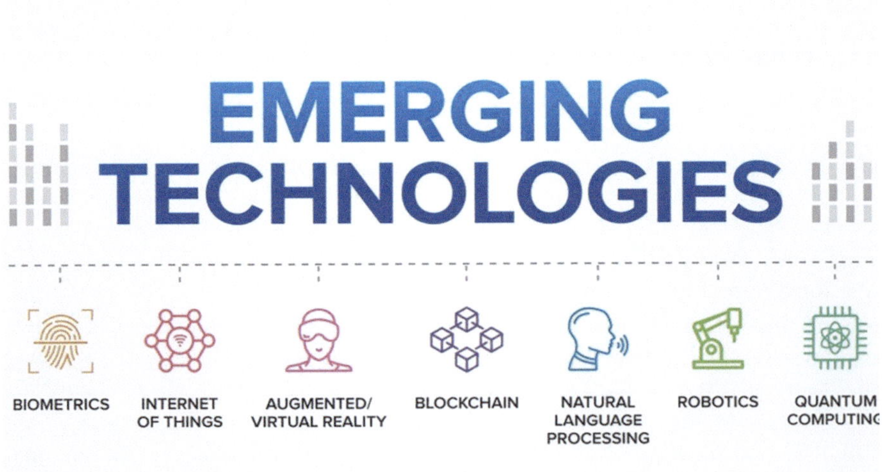

Fig. 1.4 Some emerging technologies [7]

1.3 Emerging Technologies in Architecture

The AEC sector is known for its traditional methods and slow adoption of technological advancements and innovations. In spite of this, architecture has always progressed as advances in technology and creativity push each other forward. Architectural history cycles through technological breakthroughs and explosions of creativity, with new tools and building techniques leading to a flowering of new forms. Emerging technologies are reshaping how businesses operate, innovate, and compete. Popular emerging technologies in architecture includes the following [8–10]:

1. *Building Information Modeling* (BIM): BIM is a cornerstone in AEC. It is a digital method for creating and managing detailed 3D models of buildings, including data about their parts and systems, for construction, design, and maintenance purposes. It advances the digital transformation of the AEC industry by enabling data-driven, end-to-end asset lifecycle management. BIM software has become an essential tool for architects, allowing them to create detailed digital models of buildings and structures. It lets architects capture real-world locations and landscapes and create 3D digital plans. It thus allows the work of several roles like engineers, architects, estimators, clients, and many others in a single shared process. BIM software improves project planning, scheduling, and material management, improving productivity and cost savings. Revit, ArchiCad, and Allpaln are some of the most renowned BIM-compatible software. Engineers, designers, and architects can use a BIM program to ensure their specific plans are in alignment before any cutting or welding takes place. Figure 1.5 shows different components of BIM [9].
2. *Artificial Intelligence* (AI): AI automates tasks such as researching design and code databases, design optimization, site selection, and construction management to save time. AI has started replacing much of the manual work and increasing optimization by efficiently considering all respects from weather and site conditions to experiencing virtual walkways. AI enhances the design process by analyzing factors like structural integrity, energy efficiency, material durability, and cost parameters. What is the most sustainable, material-light way to design this roof? What is the best apartment interior layout, based on thousands of previous designs? What is the right price or bid for construction materials? That is the current reality for architects and engineers, thanks to advances in artificial intelligence (AI) and machine learning, which can take oversize data sets and, with the right prompt, find efficient design solutions. These powerful logic models can take the seed of an idea and create a garden of possibilities. Engineers and architects leverage synthetic or generative AI to generate multiple design iterations that fit the client's specifications and design vocabulary.
3. *Robotics*: A robot is a machine or device programmed to perform tasks autonomously or semi-autonomously, often imitating human actions or functions. Robotics improves the productivity and efficiency of projects by automating tasks such as fabrication, assembly, and construction. Advanced

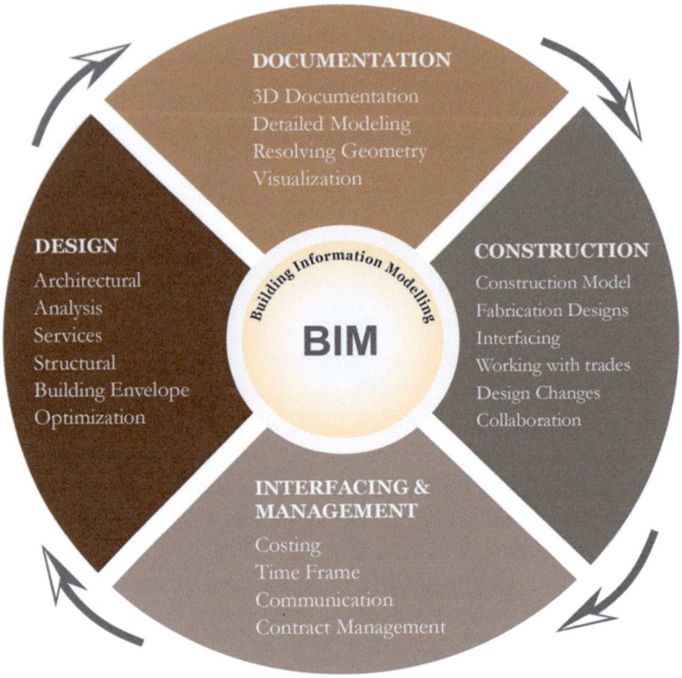

Fig. 1.5 Different components of BIM [9]

designs rests on precision construction down to the millimeter. A new generation of construction robots promises that level of accuracy and sophistication, often using laser scanners and GPS systems. Robots can perform standard design tasks, such as making models, much faster than humans. It is no sooner that more robots and humans will work together to build a project with greater efficiency. Architectural robots perform dangerous or hazardous tasks such as working at heights or confined spaces. They address the shortage of skilled labor force and improve construction site safety. Autonomous ground-based robots, including four-legged robots, are now capable of successfully navigating complex and typically cluttered construction sites. Figure 1.6 shows a typical robot [9].

4. *Drones*: Drones and UAV technologies are closely intertwined with robots and robotics, both experiencing a significant increase in interest lately as part of efforts to automate construction sites, capture geometries, detect damages. Drones, a staple in modern construction, will continue to evolve as indispensable tools for surveying and monitoring construction sites. Equipped with advanced sensors, drones will provide real-time data on project progress, site conditions and potential safety hazards. The main help of drones is fixing an object on all sides and at different heights to develop ideal 3D models. But here are more advantages of using them.

Fig. 1.6 A typical robot [9]

5. *Big Data*: Big data is definitely the big thing. It refers to the large volume of data that is being collected and processed by businesses and organizations. It has created waves in architecture in various ways. The vast volume of data that is collected and analyzed by design models can inform future building designs, aiding smart city initiatives that can improve the efficiency and sustainability of entire neighborhoods. With the greater availability of data and the Internet of things, architects are able to access more data that helps them in designing unique and more efficient buildings. Big data could be used to create "smart buildings" that are able to collect and analyze data about their inhabitants.
6. *Generative Design*: This is an emerging trend in architecture that uses artificial intelligence (AI) and machine learning algorithms to generate multiple design options based on a set of parameters. Generative design mimics how organisms function in nature. It is based on algorithms that produce a number of design variations. With the help of generative design, the work of design actually functions like evolution but is much more rapid. Generative design can help architects reduce material usage, improve structural performance, and create unique aesthetic forms. By entering constraints and goals into a script, architects can tap generative design to rapidly iterate solutions, quickly sorting through possible paths to find the right one. Dubai's Office of the Future was constructed in just 17 days, using 3D printing. Canadian startup Maket AI provides a generative AI-powered platform to create schematic designs for building floor plans. It takes environmental and other space constraints in natural language and generates various design variations. Figure 1.7 shows a typical design [2].

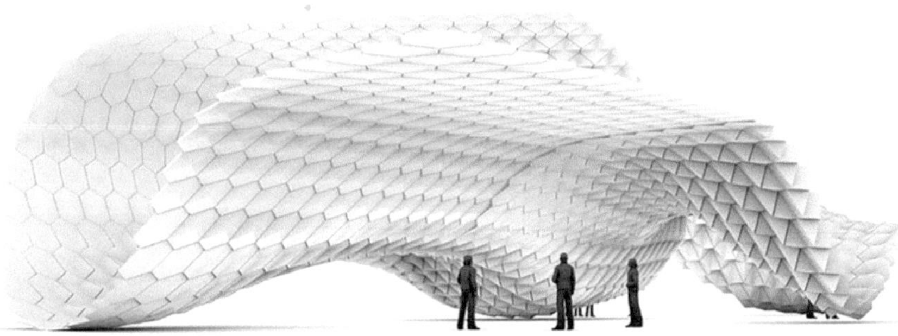

Fig. 1.7 A typical design [2]

7. *3D Printing*: It is no longer a secret that 3D printing (or additive manufacturing) has gained particular popularity among designers and architects. Creating a traditional model is time-consuming, labor-intensive, and unreasonably expensive. 3D printing is the accumulation of layers of a certain material–which can be concrete, plastic, metal, earth, and more. Additive manufacturing or 3D printing empowers architects to create unique structures that are difficult to achieve with traditional construction methods. 3D printing is a process that creates three-dimensional objects by adding material layer by layer, based on a digital model. Using 3D printing technology for architecture allows designers to invent and apply completely new and effective aesthetic forms. 3D printing technology makes it possible to reduce work completion time and improve quality. 3D printing and bonding has led to a new generation of machines that can print everything from models to entire buildings. The upsides to these massive printers are myriad, from lower material costs and less waste to dramatically faster and safer builds. 3D printing has not only enabled the building of prototype models of actual structures, but now this technology can build whole of a structure by itself. 3D printing is used to produce materials for constructing buildings, furniture, and interior details. Startups develop 3D printers that support a wide range of materials, including concrete, plastic, and recyclable materials tailored for the architectural sector. Figure 1.8 shows a building construction using 3D printing [11].
8. *Internet of Things*: The Internet of things (IoT) is a term that refers to the connection of physical objects to the internet. This can be done in a variety of ways, such as through sensors, RFID tags, or QR codes. The purpose of the IoT is to enable objects to communicate with each other and with people. A growing body of sensors, controls, and measurement tools has reshaped building operations; this suite of smart building devices is variously categorized as the Internet of things (IoT). IoT is an advanced technology for architects that links, monitors, and regulates data about the projecting object. Its main task is to competently transfer physical objects into digital ones, constantly monitoring and collecting information about these objects using sensors for data exchange.

Fig. 1.8 A building construction using 3D printing [11]

9. *Cloud Computing*: The cloud is an online service that allows you to use all systems remotely. Cloud technology refers to the use of remote servers accessed through the Internet to store, manage, and process data and applications. Today, cloud computing is a popular new technology in architecture. Solutions hosted on remote cloud servers enable architects, designers, and contractors to collaborate on projects in real time. This improves communication, reduces errors, and increases efficiency. Changing the architecture of data storage changes the way architects can access data. Remote data storage via cloud computing means that designs, digital twins, collaborative platforms, and simulations can all be hosted and shared with numerous design teams for rapid development and iteration.
10. *Immersive Technologies*: Immersive experiences allow designers, clients, and stakeholders to visualize and experience architectural designs. Architects leverage immersive technologies such as AR, VR, MR, and XR. Although classic 3D models are good at visualization, they provide a different effect than virtual and augmented reality. Designers can use virtual reality to simulate walk-throughs in the real world. VR in architecture allows one to consider how a particular solution will look with the project's other components. It allows architects to better explain their ideas and concepts to their clients in a real-life-like experience. Augmented reality adds to the real-life experience by overlaying 2D over 3D. AR layers digital images on real-world visuals, helping architects, engineers, and construction workers "see" plans applied directly to a construction site. It can also be used to conduct inspections or train workers on new building tools. Unlike virtual reality (VR), AR technology does not require connecting

headsets, glasses, or other equipment. All you need is a device camera and an app with AR functionality. Mixed Reality on the other hand allows the interaction between physical spaces and virtual spaces giving rise to concepts like vertical cities due to rapid urbanization. Extended reality (XR)—the umbrella term for both concepts, as well as mixed reality (MR)—allows ease of collaboration and communication despite physical distance. Figure 1.9 shows a typical use of VR [12].

11. *Blockchain*: Blockchain is a technology that has the potential to revolutionize the way we build. It is a distributed database that allows for secure, transparent, and tamper-proof transactions. Blockchain technology's decentralized and tamper-proof nature will bring a new level of trust and efficiency to the AEC sector. With blockchain, architects would be able to track every stage of the construction process, from design to completion. Blockchain can provide end-to-end visibility into supply chains, ensuring that products are sourced, manufactured, and delivered transparently.
12. *Architectural Apps*: Architectural apps like Revit, Grasshopper, and Rhino are being adapted by various firms worldwide and there is no stopping in more advanced versions of the existing ones or the launching of newer ones with services that are tremendously needed and are time-saving and organized.
13. *Advanced Materials*: Brick, stone, timber, and steel were the main building materials utilized until the turn of the twentieth century. Engineered materials that utilize fewer resources offer better workability over conventional materials are desirable by the AEC industry. These sustainable materials feature high durability, thermal insulation, recyclability, and endurance to withstand high loads and extreme weather. They reduce the environmental impacts of construction without compromising structural performance.

Fig. 1.9 A typical use of VR [12]

1.4 Benefits

The primary job of any design or architectural company is to make high-quality architectural models. In an era defined by rapid technological advancement, the architecture and engineering fields are undergoing transformative changes. Cloud computing and generative design enable faster prototyping. Robotics and 3D printing can rapidly assemble structures. Smart monitors and big data analysis can measure building performance and inform the next big design innovation. These digital aids allow human designers to generate and test new ideas. Technologies such as artificial intelligence (AI), virtual reality (VR), 3D printing, and cloud computing improve project quality, coordination, and implementation. Other benefits include the following [10]:

- *Visualization*: Visuals have come a long way from the days of hand-drawn sketches. The transition to digital platforms changed how architects create, communicate, and collaborate. Advances in visualization tools have made it easier to view and communicate projects with compelling depth and richness. Architects can quickly assess the impact of natural and artificial light, materials, and environmental factors on their proposals so that they can make more informed design decisions.
- *Automation*: At the intersection of design, engineering, and science, architectural technology seeks to automate and augment the design process to quicken construction, support management, and ensure high-performance of materials, systems, and structures. Artificial intelligence has the ability to automate tedious tasks, streamline operations, and enhance customer service.
- *Sustainable Design*: Sustainable and climate-positive architecture remains the top priority of the industry. Sustainable urban planning and building design is one area where new methodologies and materials could drastically reduce the carbon footprints of new projects. Technology has made it easier for architects to design sustainable buildings that minimize environmental impact. Sustainability is a driving force behind technological innovations in the AEC industry. The fundamental principles behind sustainable buildings include reducing resource consumption (energy and water efficiency), maximizing the use of renewable or recyclable resources, protecting the natural environment, and creating a healthy environment for users.
- *Renewable Architecture*: The demand for green architecture and high energy costs drive the integration of renewable energy sources in building design and construction. Architects and interior designers utilize renewables-based architecture to improve indoor air quality and natural ventilation through green roofs, biophilic design elements, etc.
- *Circular Architecture*: The architecture industry is adopting the circular economy to design and construct workflows that improve resource efficiency. Innovative circular supply chain platforms enable green construction material sourcing and increase supply chain transparency. These platforms improve regulatory compliance and incentivize greener initiatives.

- *Energy-Efficient Architecture*: Energy efficiency solutions in buildings and structures eliminate heat and energy loss as well as reduce energy and maintenance costs. This significantly curtails maintenance costs while also retaining comfort for occupants. Architecture software for energy-efficient solutions promotes passive design strategies and provides building energy simulations to improve natural lighting and building ventilation.
- *Smart Buildings*: A smart building is a structure equipped with technology and sensors that enable automated control of various systems, such as lighting, and heating. It is equipped with sensors and IoT devices that continuously collect data on energy consumption, occupancy patterns, and structural health. The Internet of things (IoT) has enabled architects to design smart buildings that are connected, intelligent, and responsive to occupants' needs. Smart buildings use sensors, algorithms, and automation systems to optimize energy consumption, air quality, and security. Technologies facilitating green building design, energy optimization, and material recycling will take center stage as the industry embraces its role in creating environmentally conscious structures.
- *Smart Homes*: These are revolutionizing the way we live now with energy-efficient systems and AI-assisted living. As populations grow and resources become more unavailable, the efficient use of limited resources becomes a necessity.
- *Smart City*: A smart city is an urban area that uses technology and data-driven solutions to improve infrastructure, services, and quality of life for its residents, while promoting sustainability and efficient resource management. Smart cities are a key factor in the consumption of resources. Urban development based on sustainability with a greater focus on renewable energy, transportation solutions, smart home energy management, and public service access are the advances that need to be realized in city technology. A typical smart city is shown in Fig. 1.10 [2].

1.5 Challenges

The construction industry has notoriously been slow to adapt to the wave of digitization spreading across various fields. The AEC sector is currently facing significant challenges and opportunities due to technological advancements. Despite the increasing interest in emerging technologies, their implementation often necessitates acquiring specific skill sets. Research indicates a noticeable gap between industry and academia in the adoption and adaptation of emerging technologies. Although emerging technologies offer immense potential for product-based companies, their successful integration requires careful planning and strategic alignment with enterprise architecture. Other challenges of emerging technologies in architecture include the following:

- *Collaboration*: Collaboration is key in architectural design, and technology has made it easier for architects to work together and share information. Cloud-based platforms enable team members to access and edit files from anywhere, promoting

Fig. 1.10 A typical smart city [2]

seamless communication and coordination. As AEC projects become increasingly global, collaboration platforms tailored to the industry will significantly advance. Architects from different parts of the world will be able to collaborate on projects using cloud-based platforms that allows them to work on the same design simultaneously. Collaboration between academia, research, and industry is crucial.

- *Education and Training*: To fully embrace technological integration, architects must invest in education and training. Professional organizations like the American Institute of Architects (AIA) offer courses in BIM, VR, and sustainable design. They provide accessible and affordable ways for architects to learn new skills and stay up-to-date with the latest technologies. Academia bears a great responsibility to remain at the forefront of the developments and educating tomorrow's AEC practitioners and researchers.
- *Aging Workforce*: The workforce in the construction sector is aging, an increase in project demand and complexity are growing, and employee productivity is falling. Even as reluctance and uncertainty plague the industry, a digital push is intensifying.

1.6 Conclusion

Now, new digital design tools are transforming how buildings are made. New digital tools have empowered architects and designers, stretching the boundaries of physics and imagination while providing the required complex calculations and visualizations. The structural ideas that underpinned skyscrapers, mosques, cathedrals, and temples evolved using increasingly sophisticated computers and digital modeling tools.

The architecture industry rapidly adopts new technologies to make smart buildings. New architecture technologies help design the buildings of tomorrow, today. They empower architects to spend more time exploring, creating, and pushing the boundaries of design. They are going to reshape the future state of application architecture and will help build faster, more secure, scalable, highly available, and innovative products for different business domains. The power of these tools has only begun to be harnessed.

Whether it is more affordable housing or ultra-modern vertical cities, architects are turning to new technologies to help them design the structures of the future. Today's developing technologies will no doubt reshape architecture and design, freeing up mental capacity to solve more complex challenges and create truly human-centered solutions. Emerging technologies increasingly influence architects' work, acting as catalysts for creativity, productivity, and communication. The integration of AI, AR, VR, BIM, robotics, blockchain, and global collaboration platforms will usher in an era where creativity meets efficiency. Exploring these emerging technologies provides valuable insights into the future direction of digital technologies in the AEC sector. More information about emerging technologies in architecture can be found in the books [13–16] and a related journal: *Automation in Construction.*

References

1. 8 Major technology advancements altering architecture, August 15, 2022. https://www.designblendz.com/blog/the-power-of-technology-in-architecture-today
2. Emerging technology trends that will help shape future of architecture. https://www.re-thinkingthefuture.com/designing-for-typologies/a3990-emerging-technology-trends-that-will-help-shape-future-of-architecture/#google_vignette
3. What is architectural technology? How innovation is changing the industry. https://www.archdaily.com/1000647/what-is-architectural-technology-how-technology-is-changing-the-industry#:~:text=Architectural%20technology%20is%20a%20vast,artificial%20intelligence%20dominate%20the%20discipline
4. M.N.O. Sadiku, P.A. Adekunte, J.O. Sadiku, Emerging technologies in architecture. Int. J. Trend Sci. Res. Dev. **9**(1), 213–223 (2025)
5. M. Halaweh, Emerging technology: what is it? J. Technol. Manag. Innov. **8**(3), 108–115 (2013)
6. N. Duggal, *Top 18 new technology trends for 2023* (2023). https://www.simplilearn.com/top-technology-trends-and-jobs-article
7. How enterprise architects integrate emerging technologies to enhance architecture. https://www.linkedin.com/pulse/how-enterprise-architects-integrate-emerging-technologies-qdgfe/
8. How new architecture technologies help design the buildings of tomorrow, today. https://www.autodesk.com/design-make/articles/technology-architecture
9. Architecture technologies: 9 Recent technological advancements in architecture that architects must know. https://www.re-thinkingthefuture.com/rtf-fresh-perspectives/a1034-9-recent-technological-advancements-in-architecture-that-architects-must-know/#google_vignette
10. Discover the top 10 architecture trends in 2023. https://www.startus-insights.com/innovators-guide/architecture-trends/

11. How new technologies are evolving to embrace sustainability in architecture. https://www.archdaily.com/991739/how-new-technologies-are-evolving-to-embrace-sustainability-in-architecture
12. 10 Architectural innovative technologies all architects should know. https://www.arch2o.com/10-architectural-innovative-technologies/#google_vignette
13. A. Ippolito, *Handbook of Research on Emerging Technologies for Architectural and Archaeological Heritage (Advances in Religious and Cultural Studies).* Information Science Reference (2016)
14. W. Quan, *Emerging Networking Architecture and Technologies: First International Conference, ICENAT 2022, Shenzhen, China, November 15–17, 2022, Proceedings (Communications in Computer and Information Science)* (Springer, 2023)
15. G. Sebestyen, C. Pollington, *New Architecture and Technology.* (Taylor & Francis, 2007)
16. Y. Xie (ed.), *Emerging Memory Technologies: Design, Architecture, and Applications* (Springer, New York, 2013)

Chapter 2
Building Information Modeling

BIM still continues to be very much at the forefront of our professional consciousness. This is hardly surprising, since BIM has been universally acknowledged as a 'disruptive technology' for the AEC industry, much more than CAD or even computing ever was, and it is causing us all to rethink our processes and identities.—Lachmi Khemlani.

2.1 Introduction

In the world of architecture, engineering, and construction (AEC), technology has continuously played a vital role in shaping the way projects are designed, planned, and executed. In the intricate realm of architectural design, innovation constantly propels the industry forward. One such innovation that has redefined the landscape is building information modeling (BIM).

The architectural drafting industry has undergone a remarkable transformation. There was a time when architects would draw by hand every excruciating detail of their design. Gradually they started using machines for the job—computer-aided designing (CAD). And now we have BIM (building information modeling), which takes the CAD concept further by embedding intelligence into the design process. In the past, construction projects were a series of individual services which built on top of each other, whereas BIM is a project with a common data repository, that provides information to all stakeholders. Figure 2.1 displays BIM project stakeholders, including architects, surveyors, engineers, contractors and the owner of the building itself [1]. BIM helps them get better connected and grow together as a community. Architects, surveyors, civil engineers, and structural engineers are professionals who use BIM to communicate information.

Architectural renderings are a valuable design and visualization asset that motivate and excites clients. A professional architectural rendering is shown in Fig. 2.2 [1], while some architects are displayed in Fig. 2.3 [1]. In addition to 3-D rendering

M. N. O. Sadiku, *Emerging Technologies in Architecture*,
https://doi.org/10.1007/978-3-031-99509-5_2

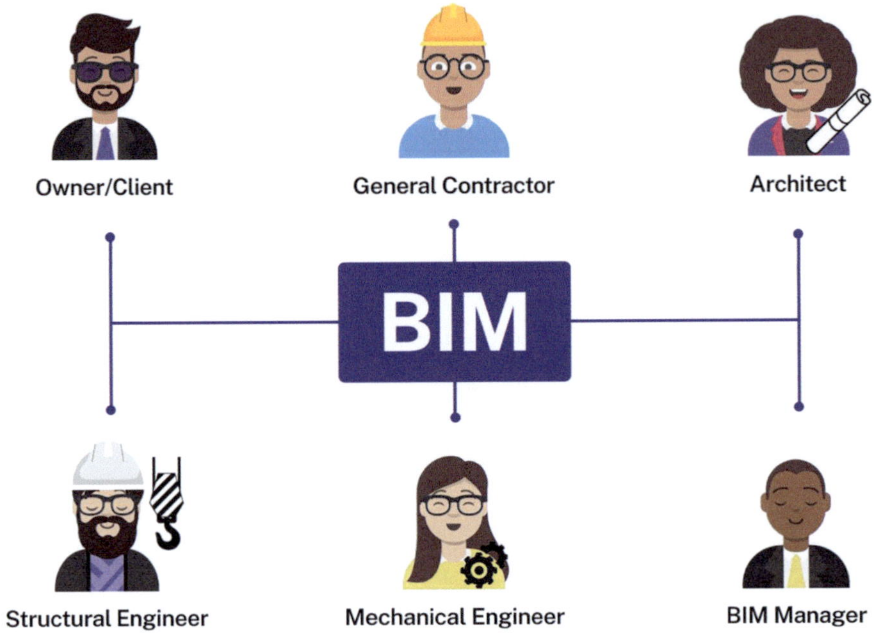

Fig. 2.1 BIM project stakeholders [1]

capabilities, BIM uses data sets to factor in time considerations (the "fourth dimension"), financial parameters (the "fifth dimension"), and asset management (the "sixth dimension"). BIM helps make sense of increasingly complex structures, such as shown in Fig. 2.4 [2]. 3D BIM models help architects, engineers, construction professionals, and facility managers better understand the entire project from start to finish.

Building information modeling (BIM) is a way of representing a built asset as a digital model. It allows architects, engineers, and other stakeholders to explore every aspect of a building–foundations, surroundings, past configurations, and even possible futures. It is the foundation of digital transformation in the architecture, engineering, and construction (AEC) industry. It has emerged as a revolutionary force in the architecture industry, transforming the way buildings are designed, constructed, and maintained. BIM, the pinnacle of architectural innovation, has ushered in a new era of design possibilities [3].

This chapter is intended to provide the use of BIM in the architecture industry. It begins with providing some background information on BIM. It discusses different levels of BIM. It provides some uses of BIM in architecture. It highlights the benefits and challenges of BIM in architecture. The last section concludes with comments.

Fig. 2.2 A professional architectural rendering [1]

Fig. 2.3 Some architects [1]

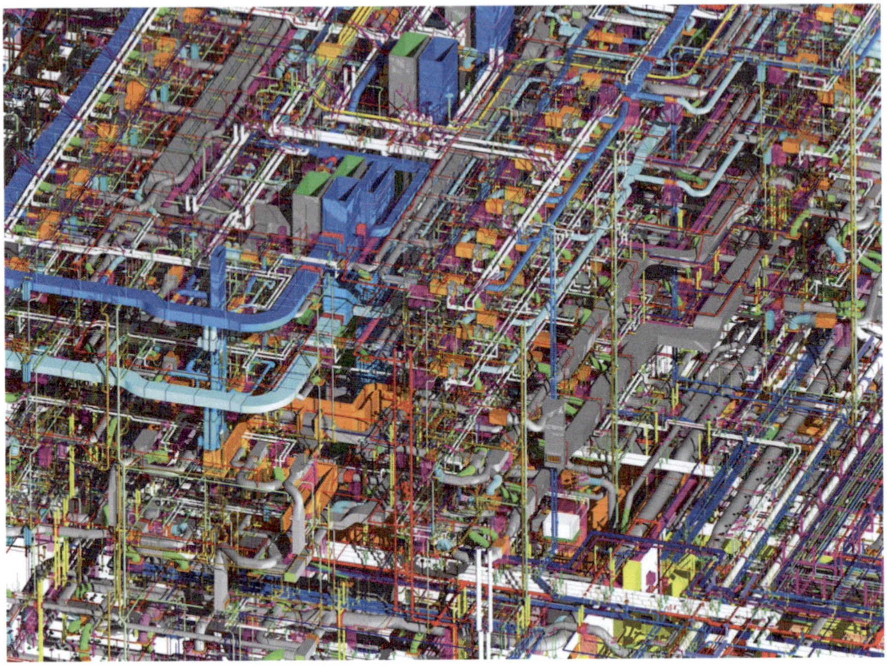

Fig. 2.4 BIM helps make sense of increasingly complex structures such as this [2]

2.2 Background on Building Information Modeling

Building information modeling (BIM) is a digital representation of physical and functional characteristics of a facility. It is the holistic process of creating and managing information for a built asset. BIM is a digital representation of the physical and functional characteristics of a building or facility. The representation can be used to simulate the construction and operation of the facility, and to provide information for decision-making throughout the project lifecycle.

BIM is a methodology that allows architects to create digital design simulations to manage all the information associated with an architectural project. It is an intelligent, 3D model-based tool that provides users with a digital representation of a facility's physical and functional characteristics. BIM is changing not only the way designs are generated, but our fundamental approach to construction in general. It allows seamless collaboration in real-time between all disciplines and all stakeholders.

Building information models (BIMs) are computer files (often but not always in proprietary formats and containing proprietary data) which can be extracted, exchanged or networked to support decision-making regarding a built asset. The main objective of creating the models is to provide builders with all relevant information about the project in a digital format. BIM software (such as Revit, Microstation, Vectorworks, Infraworks, ArchiCAD, BIM 360, Formit, Civil 3D, GBS Tekla, etc.)

allows the design team on a project to construct a 3D model that represents the one we intend to build. Two typical examples of 3D models used in BIM are presented in Fig. 2.5 [4] and Fig. 2.6 [5]. The BIM software is used by individuals, businesses and government agencies who plan, design, construct, operate and maintain buildings and diverse physical infrastructures, such as water, refuse, electricity, gas, communication utilities, roads, railways, bridges, ports and tunnels. The BIM concept envisages virtual construction of a facility prior to its actual physical construction, in order to reduce uncertainty, improve safety, work out problems, and simulate and analyze potential impacts. However, the use of BIM goes beyond the planning and design phase of a project, extending throughout the life cycle of the asset [6]. Figure 2.7 shows various components of BIM [2].

Though the 'B' in BIM stands for building, it covers all constructions in the AEC industry, from residential and industrial buildings to bridges and roads. The 'I' in BIM stands for information, referring to the integrated information within BIM models. This is its core concept—information, parameters, attributes, whatever we call it. The 'M' in BIM stands for modelling. BIM models are known to be condensed with data. Thus, BIM is not a type of software, but a holistic process for managing the building information. We use software and tools that aid the BIM process. BIM covers the entire cycle of the building, from design development to post-construction maintenance. It includes planning, designing, construction, and even operations [7].

Fig. 2.5 A typical 3D model used in BIM [4]

Fig. 2.6 Another example of 3D model used in BIM [5]

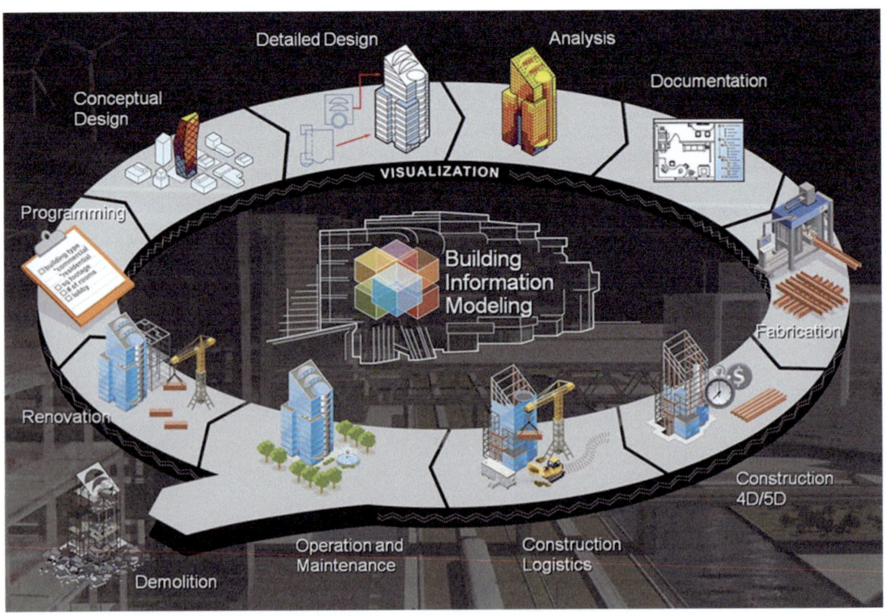

Fig. 2.7 Various components of BIM [2]

AEC professionals relied on manual drawings for decades until CAD came into existence. Traditional building design was largely reliant upon two-dimensional technical drawings (plans, elevations, sections, etc.). Building information modeling extends the three primary spatial dimensions (width, height and depth). The model can contain all the physical and functional characteristics of a project, including structure, geometry, aesthetics, materials, systems, and dynamic performance. A 3D model becomes 4D if a time component is added, and 5D when cost-related information is included. BIM enables a virtual information model to be shared by the design team (architects, landscape architects, surveyors, civil, structural and building services engineers, etc.), the main contractor and subcontractors, and the owner/operator. It enables all the stakeholders to use the same shared 3D model, from architects to surveyors, engineers, contractors, to building owners. This allows everyone to have access to the relevant information at the right moment during the design and construction of a project.

The concept of BIM has existed since the 1970s, when architects and engineers began to explore the use of computer-aided design (CAD) software to create 3D models of buildings. The terms "Building Information Model" or "Building Information Management," and "Building Information Modeling" did not become popularly used until some 10 years later. BIM only became an agreed term in the early 2000s. BIM has evolved over the past few decades from simple 2D drafting to sophisticated 3D modeling with rich data integration. Figure 2.8 shows early developments of BIM [8]. Today, BIM is an essential tool for the AEC industry. It is used throughout the entire project lifecycle, from design and construction to operation and maintenance.

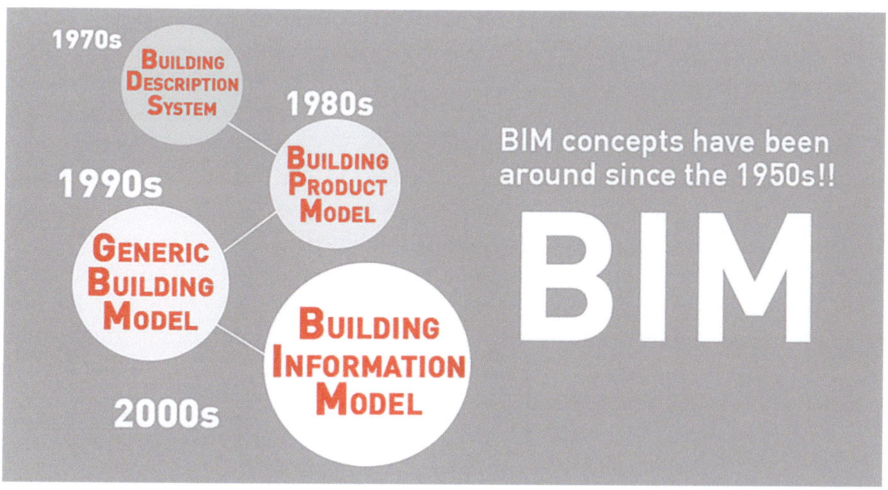

Fig. 2.8 Early developments of BIM [7]

2.3 Levels of BIM

A comprehensive BIM modeling solution combines the major elements of building information modeling software, data repository, collaboration, analysis, visualization, and project management tools [5]. Different levels of BIM can be achieved for various types of projects. Below are brief descriptions of the levels [9]:

Level 0 BIM: Paper-based drawings + zero collaboration.
Level 1 BIM: 2D construction drawings + some 3D modeling.
Level 2 BIM: Teams work in their own 3D models.
Level 3 BIM: Teams work with a shared 3D model.
Level 4 BIM: Incorporating time-related data into the BIM model.
Level 5 BIM: Integrating cost estimation and management within the BIM model.
Level 6 BIM: Including sustainability data.
Level 7 BIM: Used to manage and maintain facilities.

2.4 BIM in Architecture

BIM in architecture refers to a digital process where architects create a 3D model of a building that contains detailed information about its components, allowing for comprehensive design visualization, analysis, coordination with other disciplines like engineering, and improved project management throughout the building's lifecycle, from conception to demolition. It is a term given to the process of creating, modifying, and analyzing all the information about a building. Common BIM software used by architects include Revit, ArchiCAD, Civil 3D, Autodesk Forma, and AllPlan. They are software with which BIM is compatible. A typical Revit screenshot is shown in Fig. 2.9 [2].

Many architects, engineers, and thought leaders regard BIM as a disruptive force, changing the way building professionals design, build, and ultimately manage a built structure. BIM allows the joint work of architects, clients, builders, engineers, and other relevant actors to occur within in a single intelligent and shared process. Projects modeled in BIM can include the real products and materials that will be used to build them, incorporating their geometry, characteristics and cost into the model. Each of the elements has its own attributes and is related specifically and parametrically with the other objects of the project. If one of these objects is modified, those that depend on it will also change automatically.

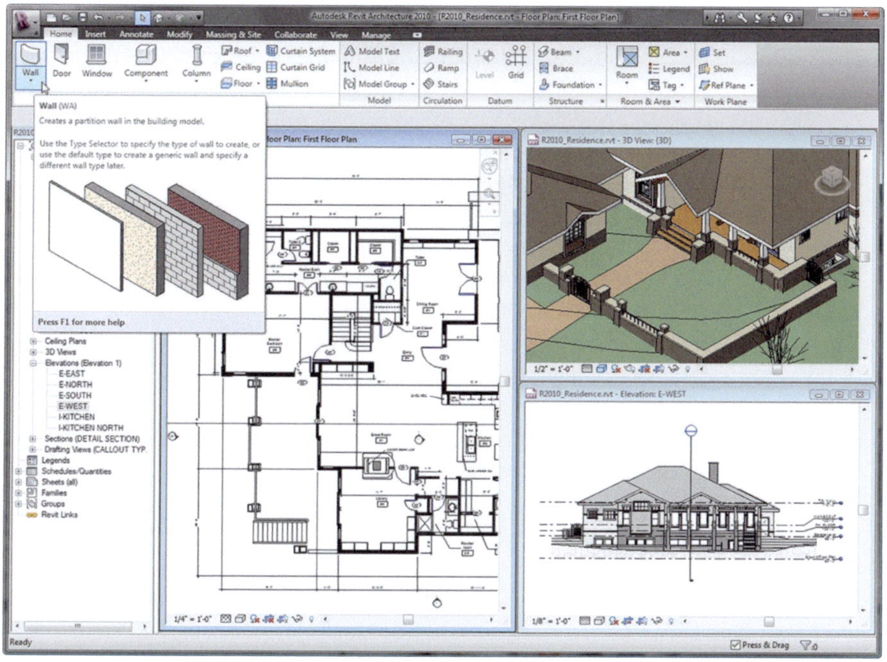

Fig. 2.9 A typical Revit screenshot [2]

2.5 Benefits

BIM spans the entire lifecycle of a construction project, from initial design to maintenance and operations. Its multidimensional approach offers unparalleled benefits across all phases of a project's lifecycle. It helps architects, engineers, and construction teams manage a project from start to finish. We are at a point where the majority of the buildings are being crafted digitally. In many nations, BIM is the government mandate for all public projects. Many companies are replacing their 2D CAD drawings with BIM files as a standard practice. Other benefits include the following [10, 11]:

- *Visualization*: One of the key advantages of BIM in architecture is its ability to enhance visualization and design exploration. Construction teams can use these visualizations to plan more effectively, identify potential risks, and make informed decisions before breaking ground. BIM provides much greater visibility into design performance and construction processes. Architects can simulate and analyze issues, while contractors can use BIM to improve work planning, coordination, and quality control. They can also create detailed 3D models that accurately represent the building's form, materials, and systems. With 3D rendering services and architectural visualization, you can actually see the entire design even before starting the construction.

- *Collaboration*: BIM enables seamless collaboration between architects, engineers, contractors, and other stakeholders by providing a shared platform to access and update project information in real-time. It serves as a shared knowledge resource, allowing all project participants to access, modify, and contribute to the building information throughout its lifecycle. It provides a simple and intuitive way for team members to work together on a project since the data is stored on a cloud-based platform. Creating a single, digital source of truth that all stakeholders can access and contribute to empowers collaboration and speeds project delivery timelines. This collaborative approach promotes better decision-making, reduces errors, and minimizes rework during the construction process.
- *Communication*: One thing that defines all BIM platforms is the blending of age-old design strategies with phenomenal new data and communication capabilities. BIM can facilitate open communication between clients and project teams, reduce costly errors, and provide greater visibility into the project as a whole. BIM not only streamlines the communication and redesign process, it also mitigates risk in the initial design phase.
- *Transparency*: BIM designs show building or infrastructure projects in their entirety. It bolsters teamwork and increases transparency throughout a project's lifespan, making each individual stakeholder in the process more accountable for their own tasks. The BIM process makes designs more transparent to all stakeholders so that they can ask more relevant questions that lead to quicker sign-offs for designs.
- *Risk Mitigation*: While clients are thrilled with the benefits of design-build construction projects, there are some drawbacks that make these efforts risky to architects and their construction contractor partners. When later design changes make costs go up, the architects and construction professionals must take a hit to their profits. The BIM process allows architects to quickly work up to more complete designs that give them realistic views of project costs.
- *Common Data Environment* (CDE): This is the central location where project members can go to access information. A CDE can reduce project bottlenecks and enable integrated data management across teams. All files, models, specifications, changes, approvals, and issues reside in a single platform to provide a high level of transparency.
- *Client Satisfaction*: The BIM process offers architects the chance to completely wow clients and secure follow-on work. By digitizing and showcasing every angle of your design, BIM helps you communicate your project vision and bring clients with you every step of the way. Architectural models and physical display models are highly effective communication tools–allowing people to visualize space, understand products and be captivated by design.
- *Civil Engineering*: BIM helps civil engineers digitally explore alternative design decisions, capture more detail, and enhance coordination. Using BIM increases accuracy, predictability, and understanding throughout project life cycles. The shared data and collaborative nature of BIM results in reduced risk, improved accuracy and constructability, and optimized designs.

- *Structural Engineering*: Structural design and engineering professionals use BIM to design, detail, document, and fabricate structural systems. BIM enables collaboration across the project team–helping to optimize designs, improve accuracy, and connect design to fabrication to deliver projects faster and more efficiently. With BIM, architects and structural engineers can get invaluable support throughout the project lifecycle in a collaborative digital environment.
- *Clash Detection*: BIM identifies potential conflicts between different building elements early in the design phase, minimizing costly construction delays and rework. Clash detection and other issue detection capabilities help teams identify and correct problems before construction when changes are easier to make. This proactive problem-solving helps reduce costly mistakes and project delays.
- *Cost Efficiency*: By streamlining processes, minimizing errors, and improving coordination, BIM can significantly reduce project costs and construction timelines. Architects can use BIM data to optimize building designs for energy efficiency, minimize resource consumption, and reduce waste during construction and operation. Early detection of potential issues and optimized material usage can reduce construction costs. Architects and engineers can automate many repetitive tasks using BIM and focus their time on design quality.
- *Improved Design*: Errors happen. Your design is subject to mistakes. By using the BIM process and the associated software, architects avoid some of the mistakes. Many BIM tools have clash detection features built into them, which automatically detect certain errors. This allows designers to correct mistakes and improve their finished design.

2.6 Challenges

While BIM offers numerous benefits, its implementation can present several challenges that need to be addressed. While the implementation of BIM may present challenges, the long-term advantages it offers are undeniable. Some challenges of BIM in architecture include the following [12, 13]:

- *Cost*: One of the primary challenges is the initial investment required for software, hardware, and training. Adopting BIM often involves significant upfront costs, which can be a barrier for smaller organizations with limited resources.
- *Interoperability*: Data management and interoperability are critical considerations. As BIM models become more complex and data-intensive, managing and maintaining this information can be a daunting task. Ensuring seamless data exchange and compatibility between different software platforms and stakeholders is crucial for successful BIM implementation.
- *Cultural Shift*: Another challenge is the cultural shift required within organizations. Adopting BIM involves changing traditional workflows, processes, and mindsets. Resistance to change and a lack of buy-in from team members can hinder the effective implementation of BIM.

- *Sustainability*: In today's environmentally conscious world, sustainable, and energy-efficient building practices are becoming increasingly important. By analyzing factors such as building orientation, material selection, and system performance, architects can make informed decisions to reduce energy consumption, minimize environmental impact, and achieve sustainability goals.
- *Standards*: To ensure interoperability and standardization across the industry, various organizations have developed BIM standards and guidelines. These standards facilitate seamless data exchange and collaboration among project stakeholders, regardless of the software or tools they use. One of the most widely adopted standards is the Industry Foundation Classes (IFC), developed by BuildingSMART International. The IFC standard is registered under ISO 16739. IFC is an open standard in the building and construction industry for digital description of building models during the BIM process. Figure 2.10 shows the various components of IFC [14]. Another important standard is the National BIM Standards (NBIMS).
- *Emerging Technologies*: Industry experts predict that adopting emerging technologies will only accelerate, driven by the increasing demand for more efficient, sustainable, and collaborative design processes. Integrating emerging technologies, such as virtual reality, augmented reality, artificial intelligence, and Internet of things, with BIM and 3D modeling will further enhance the architectural drafting landscape.

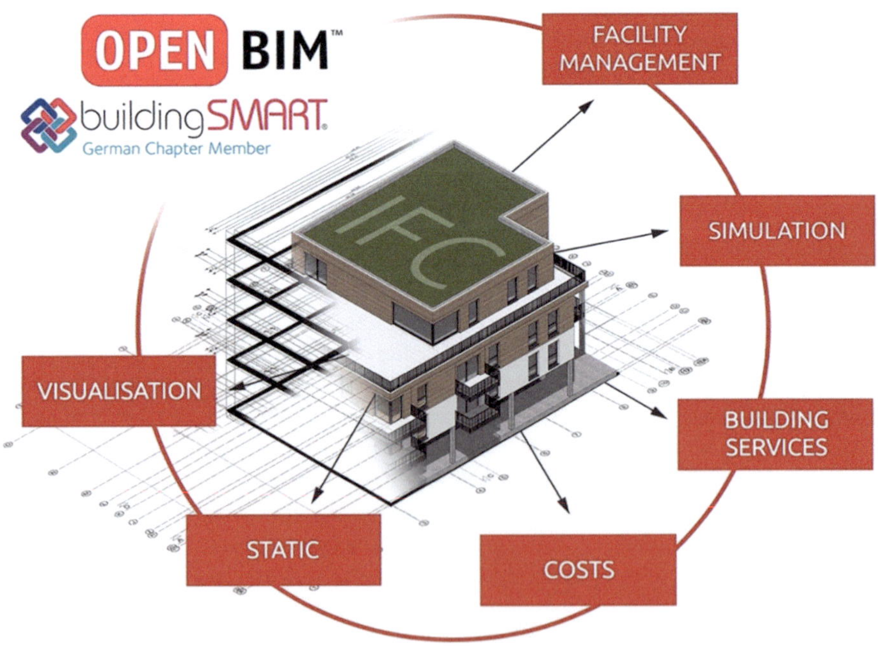

Fig. 2.10 Components of IFC [14]

- *Specialized Expertise*: As the architectural industry embraces the power of BIM and 3D modeling, the need for specialized expertise and round-the-clock support has become increasingly crucial. This has led to the rise of outsourcing as a strategic solution, allowing architectural firms to leverage global talent and resources to enhance their drafting capabilities.
- *Outsourcing*: As the architecture industry continues to evolve, the strategic advantage of outsourcing architectural drafting services becomes increasingly apparent. By partnering with experienced BIM and CAD drafting service providers, architectural companies can access the latest technologies, specialized expertise, and efficient workflows. This allows them to focus on their core competencies and deliver exceptional designs. Outsourcing can provide a range of benefits, including increased productivity, cost savings, and access to a global pool of talent.

2.7 Conclusion

The impact that evolving technology such as BIM has had on today's architectural community is too great to ignore. This cutting-edge technology offers a comprehensive and collaborative approach to project management, enabling architects, engineers, contractors, and stakeholders to work together seamlessly. BIM is well on its way to becoming a non-negotiable standard for architectural companies that want to compete for work in the future. If the construction industry wants to work toward a brighter future, embracing innovation and forward-thinking technologies clearly is the only path forward. While BIM is a game-changer for the entire construction industry, it has particular benefits for architects.

(BIM) has revolutionized the architecture, engineering, and construction (AEC) industry, offering a collaborative and efficient approach to design and construction projects. As the architecture industry continues to evolve, BIM has emerged as a transformative force, redefining the way buildings are designed, constructed, and managed. By embracing BIM, architects can unlock a wealth of benefits, including improved visualization, enhanced collaboration, cost and time savings, and sustainable building practices. With its 3D modeling capabilities, data integration, and lifecycle management features, BIM has become an indispensable tool for architects, engineers, contractors, and facility managers. More information on BIM for the architecture industry is available from the books in [15–28] and in the following related journals:

- *Construction Magazine*
- *Buildings*
- *Journal of Building Engineering*
- *Journal of Construction Engineering and Management*
- *WIT Transactions on The Built Environment*

References

1. P. Maheshwari, BIM-What is it and why is it so important for architects? https://www.novatr.com/blog/what-is-bim-for-architects
2. Young architect guide: what is building information modeling? https://architizer.com/blog/practice/tools/young-architect-guide-bim-intro/#:~:text=BIM%20facilitates%20coordination%20between%20building,problems%20and%20work%20out%20solutions
3. M.N.O. Sadiku, P.A. Adekunte, J.O. Sadiku, Building information modeling for architecture. Int. J. Trend Sci. Res. Dev. **9**(1), 224–234 (2025)
4. Building information modeling, *Wikipedia*, the free encyclopedia. https://en.wikipedia.org/wiki/Building_information_modeling
5. O. Harris, What are the key components of BIM modeling solution? https://vocal.media/education/what-are-the-key-components-of-bim-modeling-solution
6. T. Hnin, Understanding what is BIM: the complete guide 2025 (2022). https://www.novatr.com/blog/detailed-bim-guide
7. Top 5 benefits of BIM for HVAC. https://www.tcgduct.com/top-5-benefits-of-bim-for-hvac/
8. M. Sharon, Complete guide to building information modeling (BIM) (2023). https://medium.com/@matt-sharon/complete-guide-to-building-information-modeling-bim-ea0dd2cf922e
9. S. Lorek, What is BIM (building information modeling) (2022). https://constructible.trimble.com/construction-industry/what-is-bim-building-information-modeling
10. K. Forestell, Everything you need to know about building information modeling in architecture (2023). https://dozr.com/blog/everything-you-need-to-know-about-bim
11. D. Krausz, 5 Benefits of BIM for architects (2023). https://www.simscale.com/blog/bim/
12. What is BIM (building information modeling) in architecture? (2024). https://www.linkedin.com/pulse/what-bim-building-information-modeling-architecture-9gg9c/
13. M. Sharon, Role of BIM and 3D modeling in architectural drafting (2024). https://www.structuresinsider.com/post/role-of-bim-and-3d-modeling-in-architectural-drafting
14. Building Information Modeling (BIM). https://en.elitecad.eu/architecture-bim
15. R. Crotty, *The Impact of Building Information Modelling: Transforming*(Routledge, 2016)
16. S. Mordue, P. Swaddle, D. Philp, *Building Information Modeling For Dummies* (For Dummies, 2015)
17. R. Garber, *BIM Design: Realising the Creative Potential of Building Information Modelling (AD Smart)* (Wiley, 2014)
18. A.G. Garrigós, L. Mahdjoubi, P. De Wilde, *Building Information Modelling (BIM) in Design* (WIT Press, Construction and Operations III, 2019)
19. G. Elvin, *Integrated Practice in Architecture: Mastering Design-Build, Fast-Track, and Building Information Modeling* (Wiley, 2007)
20. C. Eastman, P. Teicholz, R. Sacks, *BIM Handbook: A Guide to Building Information Modeling for Owners, Managers, Designers, Engineers and Contractors* (John Wiley & Sons, 2011)
21. Y. Arayici et al. (eds.), *Heritage Building Information Modelling*(Routledge, 2017)
22. K. Kensek, D. Noble, *Building Information Modeling: BIM in Current and Future Practice* (Wiley, 2014)
23. W. Kymmell, *Building Information Modeling: Planning and Managing Construction Projects with 4D CAD and Simulations (McGraw-Hill Construction Series): Planning and ... Projects with 4D CAD and Simulations* (McGraw Hill, 2008)
24. M. De Las Nieves et al., *New Advances In Building Information Modeling And Engineering Management: Digital Innovations In Architecture, Engineering And Construction* (Springer Cham, 2023)
25. J.U. Msed, U. Isikdag (eds.), *Handbook of Research on Building Information Modeling and Construction Informatics: Concepts and Technologies* (Information Science Reference, 2009)
26. Z. Zisan, *Building Information Modeling: Transforming the AEC Industry: A Comprehensive Guide to the Future of Design, Construction, and Project Management* (Independently Published, 2024)

27. R. Deutsch, *BIM and Integrated Design: Strategies for Architectural Practice* (Wiley, 2011)
28. O. Koseoglu, Y. Arayici, *Airport Building Information Modelling* (Routledge, 2019)

Chapter 3
Artificial Intelligence in Architecture

Just as electricity is fundamental to the way we live, in the not-so-distant future, it is not hard to see how AI will become the new electricity—embedded and/or supporting just about every aspect of our life.—Daniel Eckert.

3.1 Introduction

Technology has the capacity to transform the work of architects, and as a community we should approach its offerings with openness, persistence, optimism, and skepticism.

Artificial intelligence (AI) is a rapidly evolving technology that has the potential to revolutionize various industries, including architecture. It has become a prominent player in the world of technology, revolutionizing various industries with its ability to mimic human cognitive abilities. It has the potential to expand our influence and help us be effective agents for change. AI technology uncovers extraordinary possibilities for devising innovative architectural concepts.

The term artificial intelligence (AI) is used since the 1950s of the last century. It is an aggregative term for describing when a machine mimics human cognitive functions, like problem-solving, pattern recognition, and learning. It can be regarded as the ability of a digital computer or computer-controlled robot to perform tasks commonly associated with intelligent beings. As a branch of computer science, AI drives computers to sense and learn inputs like human-being for perception, knowledge representation, reasoning, problem-solving, and planning. You may have experienced AI in a voice search with Amazon's Alexa or Apple's Siri. Figure 3.1 shows AI symbol [1].

AI enhances architectural design by automating repetitive tasks, generating multiple design iterations, improving precision, optimizing resource allocation, and ensuring sustainable building practices. AI presents a major opportunity for architects to revolutionize their workflows, make more informed decisions, and design innovative solutions. Architects should acquaint themselves with the technology while

M. N. O. Sadiku, *Emerging Technologies in Architecture*,
https://doi.org/10.1007/978-3-031-99509-5_3

Fig. 3.1 AI symbol [1]

remaining aware of its potential pitfalls. Figure 3.2 shows some architects at work [2].

Artificial intelligence (AI) encompasses a wide range of technologies that mimic human cognitive functions like problem-solving and learning from experiences. It refers to computer systems that can perform tasks traditionally done by humans. By combining principles of engineering and computer science, AI tools assist architects in getting more done in less time. AI is becoming a disruptive technology in all sectors, and architecture is no exception. It is an emerging cornerstone in the field of architectural design. It is redefining architecture, driving unprecedented efficiency, innovation, and sustainability. It is reimagining how we design, plan, and construct buildings. The integration of AI and architectural design has ushered in a new era of possibilities, challenging traditional approaches, and opening doors to unprecedented creativity, efficiency, and sustainability. AI applications in architectural design have achieved a critical mass and exploded into the mainstream of architectural imaginations [3].

This chapter examines the role of AI in the architecture industry. It begins with explaining the concepts of artificial intelligence (AI) and its subset, generative AI. It discusses AI in architecture and provides some of its applications. It highlights the benefits and challenges of AI in architecture. The last section concludes with comments.

Fig. 3.2 Some architects at work [2]

3.2 What is Artificial Intelligence?

The term "artificial intelligence" (AI) is an umbrella term John McCarthy, a computer scientist, coined in 1955 and defined as "the science and engineering of intelligent machines." It refers to the ability of a computer system to perform human tasks (such as thinking and learning) that usually can only be accomplished using human intelligence [4]. Typically, AI systems demonstrate at least some of the following human behaviors: planning, learning, reasoning, problem solving, knowledge representation, perception, speech recognition, decision-making, language translation, motion, manipulation, intelligence, and creativity.

The 10 U.S. Code § 2358 define artificial intelligence as [5]:

1. "Any artificial system that performs tasks under varying and unpredictable circumstances without significant human oversight, or that can learn from experience and improve performance when exposed to data sets.
2. An artificial system developed in computer software, physical hardware, or other context that solves tasks requiring human-like perception, cognition, planning, learning, communication, or physical action.
3. An artificial system designed to think or act like a human, including cognitive architectures and neural networks.
4. A set of techniques, including machine learning, that is designed to approximate a cognitive task.

5. An artificial system designed to act rationally, including an intelligent software agent or embodied robot that achieves goals using perception, planning, reasoning, learning, communicating, decision making, and acting."

AI provides tools creating intelligent machines which can behave like humans, think like humans, and make decisions like humans. The main goals of artificial intelligence are [6]:

1. Replicate human intelligence
2. Solve knowledge-intensive tasks
3. Make an intelligent connection of perception and action
4. Build a machine which can perform tasks that requires human intelligence
5. Create some system which can exhibit intelligent behavior, learn new things by itself, demonstrate, explain, and can advise to its user.

AI is not a single technology but a range of computational models and algorithms. The concept of AI is an umbrella term that encompasses many different technologies. AI is not a single technology but a collection of techniques that enables computer systems to perform tasks that would otherwise require human intelligence. The major disciplines in AI include [7]:

- *Expert systems*
- *Fuzzy logic*
- *Neural networks*
- *Machine learning (ML)*
- *Deep learning*
- *Natural Language Processors (NLP)*
- *Robots*

These computer-based tools or technologies have been used to achieve AI's goals. Each AI tool has its own advantages. Using a combination of these models, rather than a single model, is recommended. Figure 3.3 shows a typical expert system, while Fig. 3.4 illustrates the AI tools. These tools are gaining momentum across every industry. Analytics can be considered a core AI capability.

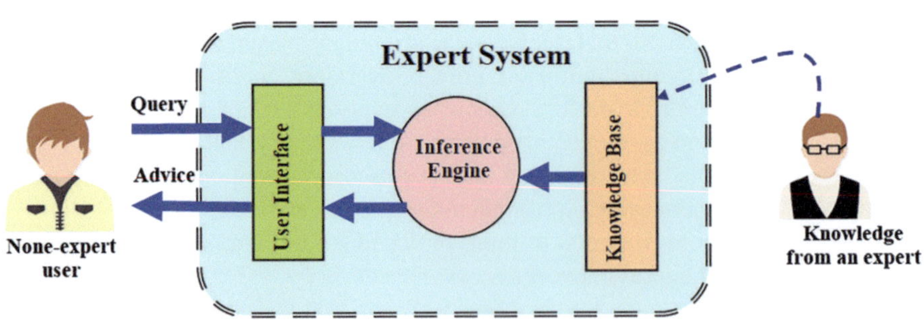

Fig. 3.3 A typical expert system

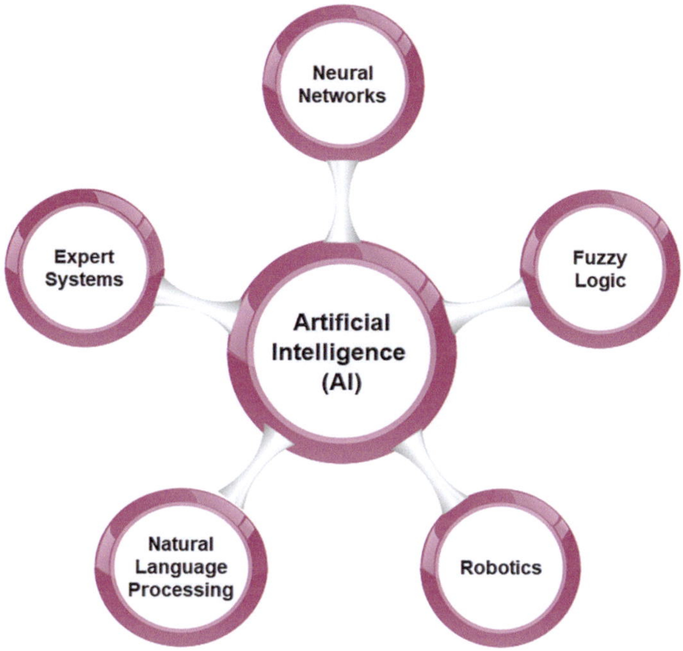

Fig. 3.4 AI tools

3.3 Generative AI

Artificial Intelligence (AI) is increasingly a part of our world and it is rapidly changing our lives. Generative AI (GenAI) is a subset of artificial intelligence that uses generative models to produce text, images, videos, or other forms of data. Generative AI (GenAI) is a term for any type of AI system capable of using generative models to create new forms of humanlike creative content, like text, images, music, audio, video and more. GenAI models include various algorithms able to learn the various patterns and structures of input training data before generating novel outputs with similar characteristics. It is essentially a narrow type and application of the broader artificial intelligence umbrella of technologies. It describes algorithms (such as ChatGPT) that can be used to create new content, including audio, code, images, text, simulations, and videos. It is specifically designed and trained to generate new content. The versatility and potential of GenAI to transform various aspects of business operations make it an attractive investment for companies across industries. GenAI uses neural networks, machine learning, deep learning models, complex algorithms, and large and varied training datasets to produce original content based on user input and how to reason in ways akin to a human brain. The technology is built on AI tools shown in Fig. 3.5 [8]. It uses neural networks to identify the patterns and structures within existing data to generate new and original content.

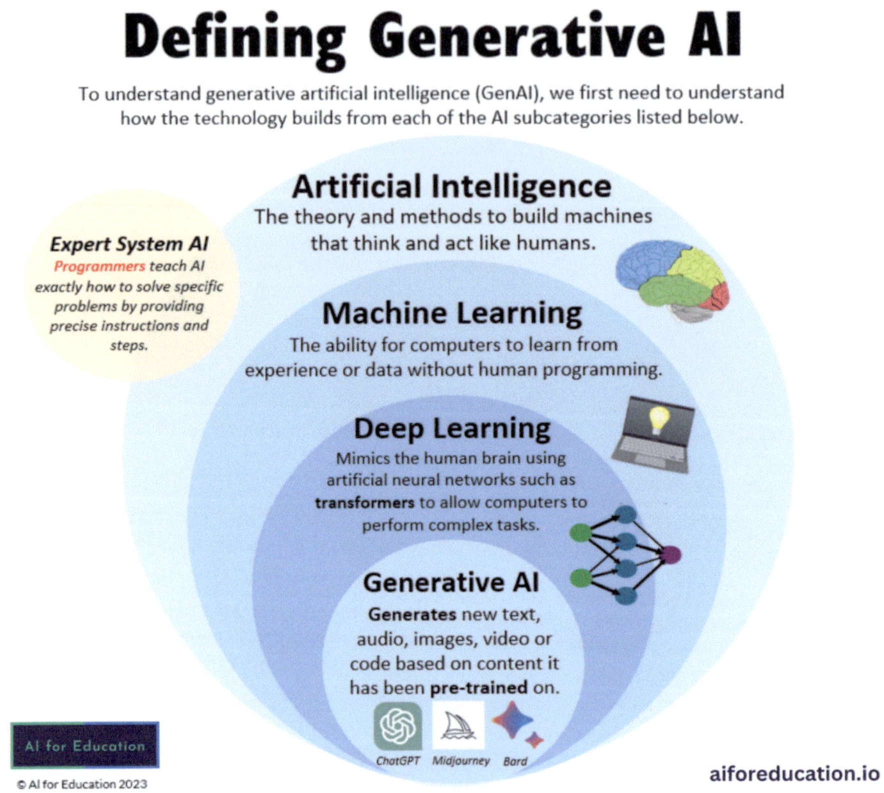

Fig. 3.5 GenAI built on AI tools listed above [8]

Generative AI can be thought of as a machine-learning model that is trained to create new data, rather than making a prediction about a specific dataset. Since its inception, the field of machine learning used both discriminative models and generative models, to model and predict data. A generative AI system is constructed by applying unsupervised machine learning or self-supervised machine learning to a data set. The most common way to train a generative AI model is to use supervised learning. Generative AI can also be trained on the motions of a robotic system to generate new trajectories for motion planning or navigation. Generative AI models are used to power chatbot products such as ChatGPT [9].

Generative AI is transforming nearly all aspects of the pharmaceutical industry, revamping the way companies operate and potentially unlocking billions of dollars in value. The pharmaceutical-operations value chain encompasses sourcing, manufacturing, quality, and the supply chain—and gen AI is expected to improve them all.

3.4 AI in Architecture

AI is essentially a multidisciplinary approach that incorporates aspects of computer science, psychology, and sophisticated mathematics to create systems capable of learning, reasoning, problem-solving, perception, and language comprehension. The impact of artificial intelligence (AI) in architecture is growing at a rapid pace. The integration of AI in architecture is enabling architects to streamline design processes. In architecture, AI is used for tasks like automating design generation, optimizing building performance, conducting real-time site monitoring, and analyzing data for sustainability.

From traditional computer-aided design (CAD) to modern building information modeling (BIM), architecture software has made quite a journey. Artificial intelligence in architecture enables engineers and architects to design, plan, and build structures more efficiently. With AI technology at hand, architects can optimize designs for sustainability and cost-efficiency and come up with never-before-seen design solutions.

AI algorithms are highly skilled at handling vast amounts of data and, when equipped with the right tools, have the potential to generate design choices. AI allows architects to brainstorm and conceptualize their ideas. AI in architecture has the potential to revolutionize the design process and create buildings that are more efficient, sustainable, and user-friendly.

With the dawn of artificial intelligence, the field of architecture is undergoing a paradigm shift that is surpassing the traditional confines of design and innovation. Artificial intelligence in architecture has various applications, from generating innovative smart city concepts to interior design visualizations. The integration of AI will significantly affect architectural design professionals in the near future. AI will not completely replace human creativity and problem-solving capabilities. AI is best used in tandem with human architects, as it can augment their skills and help make better decisions faster.

3.5 Applications of AI in Architecture

Integrating AI into your workflow can be a game-changer. Whether you use it to create images or summarize construction documents, AI clearly has multiple applications across the architecture industry. Examples of AI applications in architecture sector include automated 3D modeling, generative design, resource optimization, site analysis, virtual reality visualization, intelligent building systems, construction monitoring, data-driven design decisions, autonomous robots in construction, and AI-assisted collaboration. Common applications of AI in architecture include the following [11]:

- *Generative Design*: Generative design is an AI-driven process that creates multiple design solutions based on input parameters like dimensions, functionality, and sustainability. AI can generate hundreds of unique design options in a fraction of the time it would ordinarily take an architect. AI can rapidly produce visual designs when given specific guidelines. Generative design enables architects to explore countless design options, considering various constraints and objectives, to produce innovative and efficient solutions. Architects can leverage generative design software like Stable Diffusion, Midjourney, Dall-e 2, or Adobe Firefly to investigate different material choices and spatial designs during the initial project phases. Generative design has proven to be a fantastic tool for architects looking to create optimized, efficient, and sustainable buildings. Midjourney is everywhere right now because it can produce stunning concept design images. Figure 3.6 shows a generative design by Midjourney [12].
- *Visualization*: Imagine being able to visualize your architectural designs with an unprecedented level of realism, unlike anything that is possible before. That is exactly what AI brings to the table in the world of architectural design. By integrating AI technology with visualization tools, architects and engineers can now create animated, hyper-realistic 3D renderings of their designs. This is a game-changing tool for client presentations. Virtual reality (VR) and augmented reality (AR) are other areas in architectural visualization revolutionized by AI. Generative adversarial networks (GANs) are also making waves in architectural visualization. Figure 3.7 shows how 3D blueprints can be used to visualize designs [11].

Fig. 3.6 A generative design by Midjourney [12]

Fig. 3.7 3D blueprints can be used to visualize designs [11]

- *Image Generators*: AI image generators or modifiers can provide architects with powerful tools for exploring, optimizing, and visualizing new building designs. They are proving to be a valuable tool for architects as they allow them to explore and visualize complex design concepts with greater clarity and precision. This, in turn, is fostering a new era of creativity and enabling architects to push the boundaries of design and engineering in ways that were previously not possible. AI-based image generators can be used in the present to make the future a bit more exciting. An example of AI-based generated image is shown in Fig. 3.8 [13].
- *Pattern Recognition*: AI can analyze and identify patterns within extensive datasets, spanning architectural drawings, technical configurations, and historical projects. Currently, the process of creating a database that is easily accessible and analyzable by AI is complex, but it is anticipated to become more manageable in the future as technology advances.

Fig. 3.8 An example of AI-based generated image [13]

- *Coding*: AI can assist in coding and developing your applications and software, potentially aiding in organizing your design resources for building databases. It would be particularly beneficial if there were open-source code libraries where architects could collaborate and share information.
- *Resource Optimization*: This is one of the most important applications of AI in architecture. By analyzing critical data including energy consumption, material usage, and building performance, AI can optimize resource allocation in architectural projects while minimizing wastage. Take the use of sustainable materials, for example. AI-driven analytics can identify what materials work best with a specific design and location, reducing the need for excess materials. This not only saves costs but also helps prevent the waste of construction materials. Resource optimization also includes the application of efficient building systems, such as the use of smart sensors to auto-regulate ventilation and heating.
- *Site Analysis*: When it comes to architectural design and construction, site analysis is a critical first step. It involves evaluating various environmental and geographical factors to determine the best location for a building, which can significantly impact the success and sustainability of a project. Thanks to AI tools, architects can now optimize their site analysis process, ensuring that all critical elements are considered before breaking ground. AI algorithms can analyze geographical and environmental data to assess site conditions and identify the most suitable locations for architectural projects, taking into account factors such as topography, sun exposure, and wind patterns.

- *Construction Monitoring*: As an architect, one of the most challenging aspects of a construction project is ensuring that everything goes according to plan and stays within budget. That is where construction monitoring comes in, and AI can help streamline the process. AI-powered algorithms can analyze real-time data from construction sites to monitor progress, identify potential issues, and optimize construction workflows. AI can help track the progress of construction work, giving a real-time view of how the project is progressing. This helps in tracking the overall performance of the project and highlights any areas that require improvement.

3.6 Benefits

You can use AI to search through your knowledge base, get answers to queries, summarize meeting notes, and more. By embracing AI, you stand out from the crowd, making your designs stand out as more unique, innovative, and efficient. Other benefits of integrating AI in architecture include the following [11, 12]:

- *Creativity*: AI can serve as a powerful tool for unlocking new levels of creativity in architecture. It simulates multiple designs and variations based on specific, pre-defined parameters. This provides architects with a variety of design possibilities and iterations. As a result, architects can pursue creative ideas that push the boundaries of architectural design.
- *Automation*: AI allows automating repetitive tasks such as adapting to changes or variations, conducting compliance checks, and generating prototypes of 3D models. This unlocks room for human expertise as architects pursue their creative or strategic interests. Generative design tools auto-populate numerous design variations depending on input parameters.
- *Collaboration*: Collaboration is key to the success of any architectural project. However, the traditional methods of communication and data sharing can be time-consuming and prone to errors. AI facilitates seamless collaboration among architects, engineers, and other stakeholders. Real-time data sharing and collaborative platforms are powered by AI enhance communication throughout the design process. This integrated approach reduces errors, enhances coordination, and ultimately leads to more successful and efficient projects. An example of AI-assisted collaboration is the use of virtual assistants in architectural companies.
- *Sustainability*: AI can help architects design energy-efficient and sustainable buildings. By analyzing factors such as climate data, building orientation, material, and energy consumption patterns, AI algorithms can recommend design modifications that make building energy use more efficient and reduce its environmental impact. Through smart material selection and energy optimization, AI promotes a more sustainable future in architecture. Figure 3.9 shows a typical example of smart and sustainable design [14].

Fig. 3.9 A typical example of smart and sustainable design [14]

- *Data Summarization*: AI can assist in condensing information from books and reports, making it easier for you to locate essential details swiftly while still having the source available for review.
- *Building Maintenance*: AI tools have been developed to analyze video feeds and detect weaknesses in areas such as road conditions and the exteriors of existing buildings, which enables early detection of required maintenance for infrastructure.
- *Building Materials*: Until recent times, construction material selection relied on experience and intuition. AI algorithms eliminates guesswork from this design process through their impressive research abilities. It is a powerful tool for research as it pores through large data repositories of building materials in almost no time. It can help identify innovative materials, as shown in Fig. 3.10 [14].
- *Emerging Technologies*: AI can analyze BIM (building information modeling) data, spot potential conflicts or clashes, optimize schedules, and aid in project management activities, thus enhancing coordination and diminishing errors in the construction process. Architects can harness the power of AI-driven virtual reality (VR) and augmented reality (AR) technologies to create visualizations and presentations of their designs.
- *Efficiency*: The use of AI in architecture allows for the collection and interpretation of real-time data. This equips architects and engineers with the insights necessary to devise solutions for optimal operational efficiency. By automating

Fig. 3.10 Some innovative materials [14]

time-consuming tasks, AI gives you more time to focus on conceptualization and design. Integrating AI in schematic planning leads to highly efficient and optimized layouts.

- *Productivity*: AI significantly boosts efficiency and productivity in architecture by automating repetitive and time-consuming tasks. Tools powered by AI can handle tasks like initial floor plan drafting, compliance checks with building codes, and running environmental simulations much faster than traditional methods.
- *Predictive Maintenance*: AI plays a critical role in predictive maintenance and risk management. AI does not just react to present conditions; it is efficient at predicting future circumstances too. This comes to the fore with predictive maintenance, where AI can foresee equipment or system issues before they escalate, enabling proactive preventive care. Such analyses do not just reduce repair costs and improve system longevity, but also ensure uninterrupted comfort for the building's users.
- *Cost Savings*: AI in architecture delivers substantial cost savings by streamlining both the design and construction processes. By automating tasks such as drafting, compliance checks, and material calculations, AI minimizes labor costs and shortens project timelines. AI tools reduce project costs and timelines by optimizing the design process and automating tasks. Such an approach ensures project delivering on time and within budget. AI also optimizes the use of materials, reducing waste, and cutting down on storage expenses through efficient supply chain management, ensuring that materials are delivered in the right quantities and at the right time.

- *Site Safety*: AI enhances construction safety by monitoring site conditions in real-time, detecting safety risks such as worker proximity to machinery, and providing alerts to supervisors for timely interventions. AI can mitigate health and safety risks.

3.7 Challenges

While the advantages of using AI in architecture are evident in enhanced creativity, efficiency, and sustainability, architects must also tread carefully to address some challenges such as over-reliance, data privacy and security concerns, and ethical considerations. When a technology is this revolutionary, it should come as no surprise when concerns are raised. The loss of creativity, job displacement, and ethical dilemmas are just a few examples of concerns about the adoption of AI in architecture. Other challenges include the high cost of AI technology, the learning curve for integrating AI into workflows, and the potential for data accuracy issues that could impact design outcomes. More challenges of AI in architecture include the following [15]:

- *Resource Management*: This is singularly the toughest part of any construction process. Everything, from software platforms to design concepts to material and labor at the construction site, constitutes resources. AI platforms help ensure the optimum use of resources.
- *Replacing Architects*: AI has the potential to automate up to 37% of the tasks typically carried out by architects and engineers. This automation is likely to target routine and less creative tasks, allowing professionals to concentrate on more strategic and imaginative aspects of their work. Just as Revit and 3D software did not replace architects but only transformed their workflows, the same principle holds for AI tools.
- *Job Losses*: While AI is unlikely to cause mass job losses for architects shortly, it will fundamentally change the nature of their profession. To effectively embrace new tools and evolving knowledge, architects must remain current and continually update their skill sets.
- *Risk Mitigation*: The construction industry is naturally complex with inherent risks. You run the risk of delays, inflationary pressures, skill shortages, scope creep, safety hazards, natural calamity—the list goes on! In the construction and architectural industries, managing risks is a critical yet complex task. You can use AI to manage and mitigate risks in the architectural landscape. AI's ability to suggest unconventional construction methods or materials may pose liability issues, making it essential for humans to remain in control of critical decisions.
- *Loss of Creativity*: One major concern is the potential loss of human intuition and creativity. While AI can enhance creativity by generating novel design alternatives, there is a risk of over-reliance. Depending too heavily on algorithms may stifle the architect's innate creativity and intuition, potentially resulting in designs that lack the human touch.

- *Data Accuracy*: AI's effectiveness depends on the quality of the data it analyzes. Inaccurate or biased data can result in flawed designs, raising concerns about the integrity of AI-assisted decisions. Ensuring clean, reliable data inputs is essential to prevent costly mistakes.
- *Black Box*: Black box AI is the lack of visibility into how an algorithm reaches its conclusion. This can lead to unintended consequences as algorithms tend towards optimization based on training data instead of desired outcomes and can produce results with no explainable logic behind them. An architect may not be able to decipher what exactly the model is doing.
- *Privacy and Security Concerns*: The extensive use of AI in architectural design involves handling large volumes of sensitive data. Ensuring the privacy and security of this data becomes paramount, as breaches could compromise the integrity of projects, intellectual property, and the personal information of stakeholders.
- *Initial Costs*: The high upfront investment required for AI technologies and infrastructure is another barrier, especially for smaller companies. Implementing AI tools in architectural firms requires a financial investment. The initial costs of acquiring and integrating AI systems, along with the time and resources needed for training staff, can pose challenges for smaller practices.
- *Ethical Considerations*: As AI algorithms influence design decisions, ethical concerns arise. Questions regarding bias in algorithms, accountability for design choices made by AI, and the ethical implications of AI-generated designs need careful consideration..
- *Inclusive Architecture*: The concept of inclusive architecture has gained prominence as the built environment evolves to reflect and address the diverse needs of humanity. This approach prioritizes empathy, accessibility, and equity, striving to create spaces that resonate with individuals across a spectrum of demographics, abilities, and cultural contexts. Inclusive architecture fosters connections, embraces diversity, and ensures that physical spaces contribute to collective well-being.

3.8 Conclusion

The architectural landscape is evolving rapidly, and architects are gradually gaining a clearer comprehension of AI's capabilities and boundaries. Architects must continue to embrace AI not only as an efficient tool but also as a creativity ally.

The impact of AI in the field of architecture has been remarkable, leading to an improved workflow efficiency, optimized resource utilization, and sustainable designs. As the technology evolves, we can expect further enhancements that will expand AI's reach and applicability in architecture. The future of AI in architecture and design holds incredible promise, pushing the boundaries of architecture even further. More information on artificial intelligence in the architecture industry is available from the books in [16–30] and the following related journals:

- *The AI Journal*
- *AI Magazine*
- *Building Magazine*
- *Journal of Intelligence*
- *Journal of Building Engineering*
- *Journal of Construction Engineering and Management*
- *Automation in Construction*

References

1. The rise of artificial intelligence in construction industry (2022). https://gharpedia.com/blog/artificial-intelligence-in-construction/
2. A. Ahramovich, Artificial intelligence in architecture: 10 Use cases and top technologies (2023). https://www.itransition.com/ai/architecture
3. M.N.O. Sadiku, O.D. Olaleye, J.O. Sadiku, Artificial intelligence in architecture. Int. J. Trend Res. Dev. **12**(1), 95–101 (2025)
4. M.N.O. Sadiku, Artificial intelligence. IEEE Potentials 35–39 (1989)
5. Artificial intelligence (AI). https://www.law.cornell.edu/wex/artificial_intelligence_(ai)
6. Artificial intelligence tutorial. https://www.javatpoint.com/artificial-intelligence-tutorial
7. D. Quinby, Artificial intelligence and the future of travel (2017). https://www.phocuswright.com/Travel-Research/Research-Updates/2017/Artificial-Intelligence-and-the-Future-of-Travel
8. Generative AI explainer, Unknown Source
9. M.N.O. Sadiku, P.A. Adekunte, J.O. Sadiku, Generative artificial intelligence. Int. J. Trend Sci. Res. Dev. **8**(6), 561–570 (2024)
10. Will artificial intelligence replace architects? https://www.archdaily.com/1007802/will-artificial-intelligence-replace-architects
11. S. Lucas, 10 AI applications in architecture and design (2024). https://futurearchi.blog/en/ai-applications-architecture-design/
12. AI in architecture: Boosting design efficiency for a competitive edge (2023). https://www.e-zigurat.com/en/blog/ai-in-architecture-guide/
13. P. Sharma, 15 best AI image generators you should try (2023). https://parametric-architecture.com/15-best-ai-image-generators-you-should-try/?srsltid=AfmBOorGhMwruuWq18cMcz6m391fD
14. M. Pavitra, Building with technology: a guide on how to use ai in architecture (2024). https://clickup.com/blog/how-to-use-ai-in-architecture/
15. H. Syed, Navigating the frontier: the pros and cons of AI in architectural design (2024). https://www.linkedin.com/pulse/navigating-frontier-pros-cons-ai-architectural-design-hamza-syed-lrmsf/
16. M. Sourek, *Artificial Intelligence in Architecture and the Built Environment* (CRC Press, Boca Raton, FL, 2025)
17. S. Chaillou, *Artificial Intelligence and Architecture: From Research to Practice.* Birkhäuser, 2nd edn (2025)
18. H.R. Garcia, *The Impact of Artificial Intelligence on Architecture and Architectural Visualization* (Independently Published, 2023)
19. M. del Campo (ed.), *Artificial Intelligence in Architecture* (Wiley, 2024)
20. N. Leach, *Architecture in the Age of Artificial Intelligence: An Introduction to AI for Architects* (Bloomsbury, 2025)

21. G. Gallo, F. Wirz (eds.), *The role of Artificial Intelligence in architectural design: conversation with designers and researchers, Proceedings of S. Arch 2020, the 7th International Conference On Architecture and Built Environment* (2020)
22. F. Jacobus et al., *Artificial Intelligent Architecture: New Paradigms in Architectural Practice and Production* (ORO Editions, 2023)
23. P. Bernstein, *Machine Learning: Architecture in the Age of Artificial Intelligence* (RIBA Publishing, 2022)
24. I. As, P. Basu (eds.), *The Routledge Companion to Artificial Intelligence in Architecture* (Routledge, 2023)
25. S. Blomhagen, *How to Use Artificial Intelligence In Architecture: A Practical Guide For Architects* (Independently Published, 2024)
26. D.R. Martinez, B.M. Kifle, *Artificial Intelligence: A Systems Approach from Architecture Principles to Deployment (MIT Lincoln Laboratory Series)* (The MIT Press, 2024)
27. J. Yao et al. (eds.), *Architectural Intelligence: Selected Papers from the 1st International Conference on Computational Design and Robotic Fabrication (CDRF 2019)* (Springer, 2020)
28. M. del Campo, *Neural Architecture: Design and Artificial Intelligence* (ORO Editions, 2022)
29. M.W. Steenson, *Architectural Intelligence: How Designers and Architects Created the Digital Landscape* (The MIT Press, 2022)
30. M. del Campo, N. Leach (eds.), *Machine Hallucinations: Architecture and Artificial Intelligence* (Wiley, 2022)

Chapter 4
Generative Design in Architecture

As designers, we have such power to make design change the world. We really need to make better use of our platform and do stuff that is useful.—Yinka Ilori.

4.1 Introduction

Innovations in technology have continually pushed the boundaries of what is achievable in the realm of architecture. From building information modelling (BIM) to augmented reality and the Internet of things (IoT), it is no secret that technology has transformed the way architects work and revolutionized the industry. In today's rapidly evolving architectural landscape, the fusion of technology and creativity has ushered in a new era of design possibilities. At the forefront of this revolution are building information modeling (BIM) and generative design. BIM encompasses a digital representation of physical and functional characteristics of facilities. Generative design is a design exploration process that utilizes algorithms to generate and evaluate numerous design alternatives based on specified goals and constraints, allowing architects to discover innovative solutions that may not be immediately apparent through conventional design methods. While BIM focuses on creating a digital representation of a building's physical and functional properties, generative design explores and evaluates design alternatives based on predefined criteria. BIM and generative design complement each other in the design process. BIM provides the data infrastructure for generative design, enabling seamless integration and collaboration between the two approaches [1].

Generative design is a new innovation shaping the architectural industry. It is a strategy where the architect or user inputs data into their chosen software. Figure 4.1 shows an architect at work [2]. Generative design is a design method that uses algorithms and artificial intelligence (AI) to generate multiple design solutions based on specific requirements and constraints. It is the process of inputting design goals, materials, cost constraints, and other data points into software that evaluates all possible solutions. It is an iterative design process used in architecture and manufacturing. The

M. N. O. Sadiku, *Emerging Technologies in Architecture*,
https://doi.org/10.1007/978-3-031-99509-5_4

Fig. 4.1 An architect at work [2]

power of this approach lies in its ability to create countless alternatives, surpassing the capabilities of the human mind. Generative design in architecture is a process that uses algorithms to generate design options based on user-defined criteria. Architects have been using digital documents for decades, from CAD drawings to spreadsheets. Generative design has ushered in a new era of architectural innovation, empowering designers to envision and construct buildings that were once considered beyond the realm of imagination [3].

This chapter examines the various uses of generative design in architecture. It begins with explaining the concept of generative design. It discusses generative design in architecture and provides some of its applications. It highlights the benefits and challenges of generative design in architecture. The last section concludes with comments.

4.2 What is Generative Design?

For decades, traditional design methods have been plagued by repetition and rigidity. Designers spend weeks or months developing, testing, adjusting, and refining designs, with no guarantee of optimal results. Generative design has emerged as a revolutionary solution, utilizing artificial intelligence and advanced algorithms to automate and optimize the design process. It is an iterative design process that uses software to generate outputs that fulfill a set of constraints iteratively adjusted by a designer. This process often involves multiple iterations until the product meets the specified technical parameters [4]. It uses software to generate outputs that fulfill a set of constraints iteratively adjusted by a designer. Examples of software that provide an option for generative design are Autodesk Revit, Infurnia, Creo, and Rhino. For example, Fig. 4.2 shows the generative design by Autodesk [5]. These software utilize AI and machine learning algorithms to learn from each iteration and continually improve the output. One may regard the software as an assistant that helps with creating, testing, and evaluating options. They have emerged as a game-changer for professionals looking to advance their careers.

Generative design is a new design method that overcomes the limitations of traditional design. Using algorithms and artificial intelligence, generative design software automatically generates thousands of different design solutions. Figure 4.3 shows the traditional component design as an iterative process [4], while Fig. 4.4 shows the generative design [4]. The iterative process provided by the computer software enables the trial-and-error approach in design, and involves architects interfering

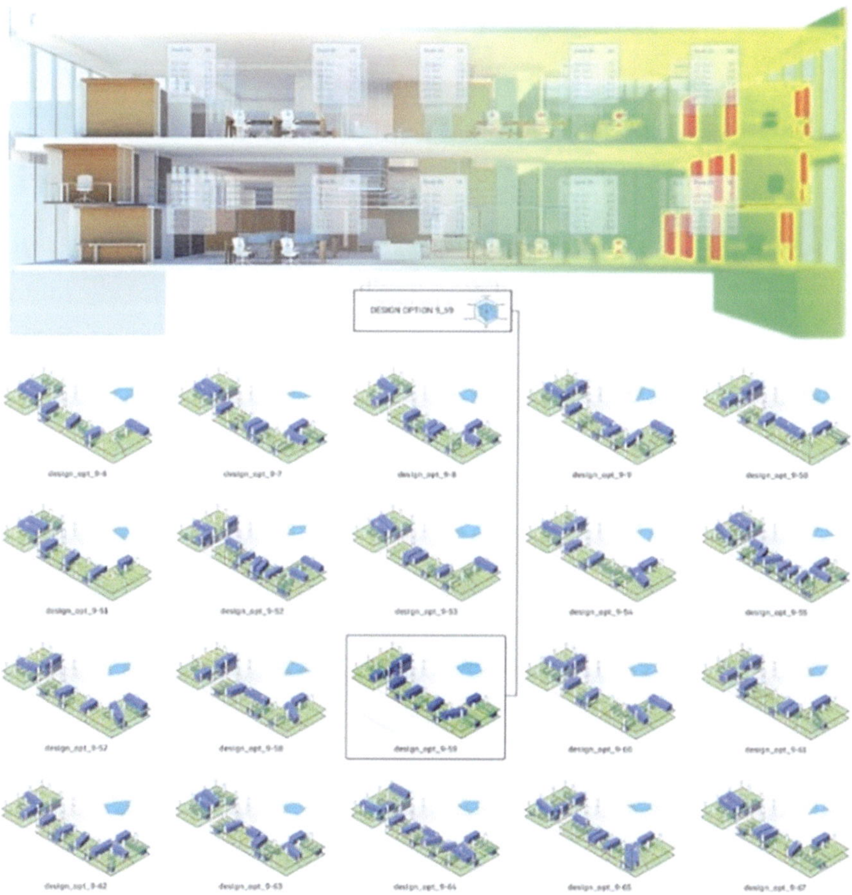

Fig. 4.2 Generative design by Autodesk [5]

with the optimisation process. Generative design is not just about creating a multitude of options. It is about refining and optimizing these options to reach the most effective solution that best meets the defined parameters and constraints.

Generative design can enhance the existing design process and to make it more efficient and effective for architects and all those involved. Generative design in architecture is a method that uses algorithms and artificial intelligence to analyze, explore, and generate a series of design possibilities.

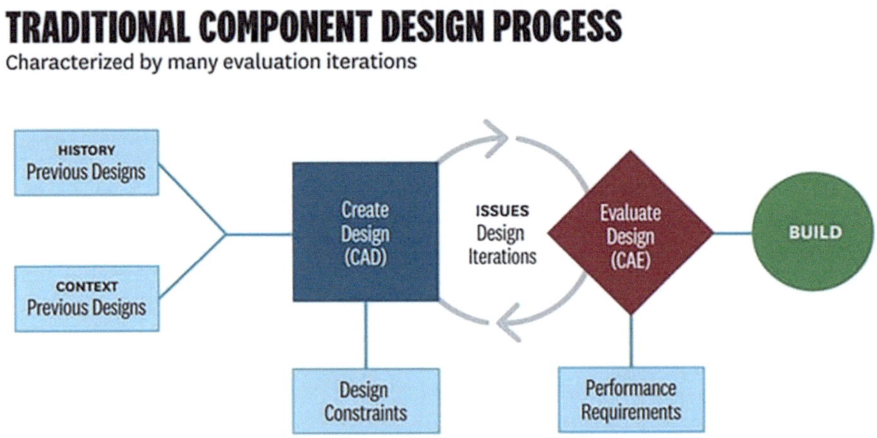

Fig. 4.3 Traditional component design as an iterative process [4]

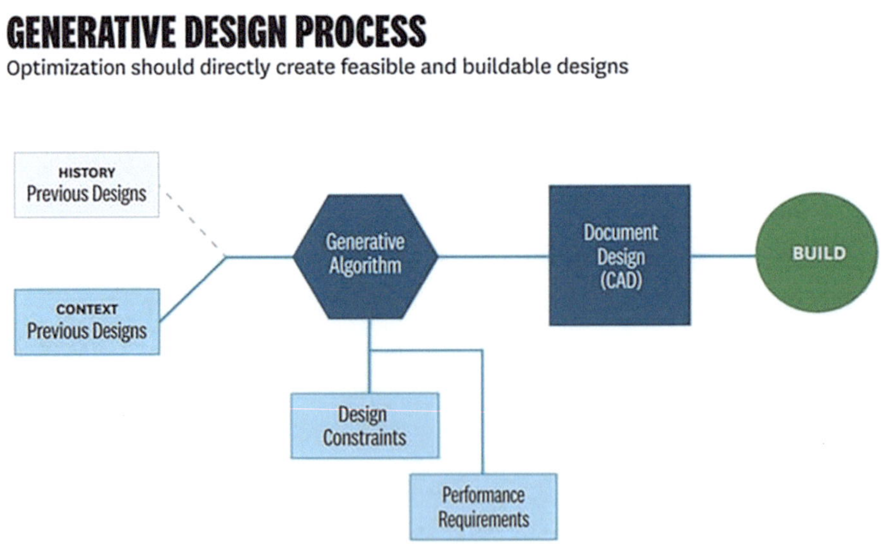

Fig. 4.4 Generative design [4]

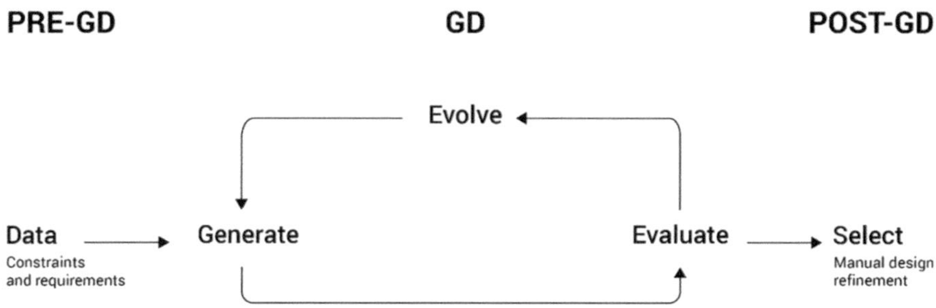

Fig. 4.5 Generative design process has three overall stages [5]

Artificial intelligence (AI) and machine learning (ML) are at the core of the generative design process. They play a crucial role in generative design by automating the design process. AI can quickly generate and evaluate different design options. Generative design integrates AI into the design process by using metaheuristic search algorithms to discover novel and high-performing results within a given design system.

The system works by iterating through various options and optimizing them to find the best design. The designer inputs their requirements into the generative design tool. Then, the tool will generate different designs according to the designer's requirements and local regulations. As shown in Fig. 4.5, the process has three overall stages [5]. They are:

1. *Pre-Generative Design*: The preliminary stage is called pre-generative design (Pre-GD). At this stage, you have to define goals, identify important parameters that are important in achieving them and specify constraints. This data might include design goals, time or cost constraints, materials and other relevant information. These constitute a set of data that are used to kick start the main generative design process. The more precise data gathered and input into the generative process the more efficient and fine-tuned results will be achieved.

2. *Generative Design*: The second stage or the main stage is where computer algorithms come into play. Provided with the data, it starts a cyclical process of generative design (GD), which can be divided into three stages itself.

- Generate: From a seed given by the human operator and using the defined parameters, the computer algorithm creates a variety of designs.
- Evaluate: The computer analyzes the results and calculates their compatibility with the requirements defined at the preliminary stage.
- Evolve: The algorithm decides the direction of a new cycle of design, based on the most compatible out of the evaluated options, using that option as a new seed to begin the whole process all over again.

This process repeats the cycle several times before moving on to the next stage.

3. *Post-Generative Design*: The third stage or the finalizing stage is called post-generative design (Post-GD). At this stage, the designer and stakeholders can explore

and choose between a final set of proposed designs produced by the process, and then the architects and professionals can develop and manually refine the chosen option further and ensure constraints and requirements are met. It is composed of two key steps: selection and refinement.

4.3 Generative Design in Architecture

Architectural design is an iterative process. It requires multiple trials and errors to arrive at the best possible solution. Traditionally, sketchbooks, pencils, pens, and markers have been architects' best friends throughout their design journey, from ideation to realization. This is typically illustrated in Fig. 4.6, where architects use traditional tools [6]. These conventional tools are time-consuming and do not offer enough agility to experiment with design. The use of computers has set the stage for a digital revolution in architecture. Next-generation design technology, such as generative design, is taking the architecture, engineering, and construction (AEC) industry by storm.

Generative design in architecture is an iterative design process that enables architects to explore a wider solution space with more possibility and creativity. It is a new method of design that enables you to discover unexpected novel designs and

Fig. 4.6 Traditional architects [6]

navigate trade-offs between high-performing designs, sketch constraints, and goals rather than form and co-design between human and computer. It empowers architects to explore unconventional design solutions, optimize building performance, and respond to complex design challenges with agility and creativity. Compared with traditional top-down design approach, generative design can address design problems efficiently, by using a bottom-up paradigm that uses parametric defined rules to generate complex solutions.

Generative design is disruptive; it changes the role of architects and the entire design methodology. The architect's role shifts from creating designs to setting the parameters and constraints for the generative model. The architect will use computers not just to describe buildings, but cocreate them. Such methodology introduces new ways to think, make, and produce architecture, streamlining collaboration efforts between designers and clients while unlocking the creative power of artificial intelligence for designers and engineers.

4.4 Applications of Generative Design in Architecture

Generative design is not only a powerful design tool but also offers diverse applications in numerous fields such as aerospace, aviation, automotive industry, manufacturing, construction, and architecture. Within architecture and construction, generative design helps optimize building structures to ensure sustainability and cost-effectiveness. By leveraging advanced computational tools and techniques, architects can push the boundaries of architectural expression and create iconic structures that redefine the urban landscape. Generative design allows architects and builders to use data to create new and innovative solutions to common building roadblocks. Common areas of applications include the following:

- *Building Design*: Buildings and bridges designed using this method often possess complex structures while maintaining durability and efficient material utilization. Generative design tools like Autodesk's Project Refinery have been used to optimize building designs for sustainability, resulting in reduced environmental impact and resource consumption. By integrating environmental performance metrics into the design process, architects can create buildings that are not only aesthetically compelling but also environmentally responsible and resilient. Generative design allows architects and builders to use data to create new and innovative solutions to common building roadblocks. Figure 4.7 shows a typical building design [4].
- *Space Design*: In architecture, drawing is a technical and artistic expression that involves creating visual representations using various analog instruments. While drawing remains relevant and current in practice today, efforts have been made to carry out architectural tasks and studies more efficiently. The emergence of computational tools, such as computer-aided drafting (CAD), has revolutionized the workflow by leveraging the advantages offered by computers. Architects can

Fig. 4.7 A typical building design [4]

now play a more direct and creative role in the design process, reducing their reliance on time-consuming drawing and repetitive tasks. A typical space design is shown in Fig. 4.8 [6].

- *Additive Manufacturing*: Combining generative design technology with advancements in additive manufacturing (or 3D printing) has the potential to fundamentally change how we think about architecture. Instead of designing building envelopes made up of separate layers for heating, ventilation, passive solar gain, and other necessities, these needs could be accounted for within a single, complex "skin" that possesses qualities that mimic biological organisms. The possibilities are startling.
- *Vertical Cities*: As one of the tallest skyscrapers in the world, the Shanghai Tower in China showcases the immense possibilities of generative design in creating sustainable and resilient vertical cities. The tower used generative design algorithms to optimize the building's shape for wind resistance and energy efficiency. The tower's iconic twisting form not only enhances its visual appeal but also gives the building structural stability. The tower is shown in Fig. 4.9 [7].

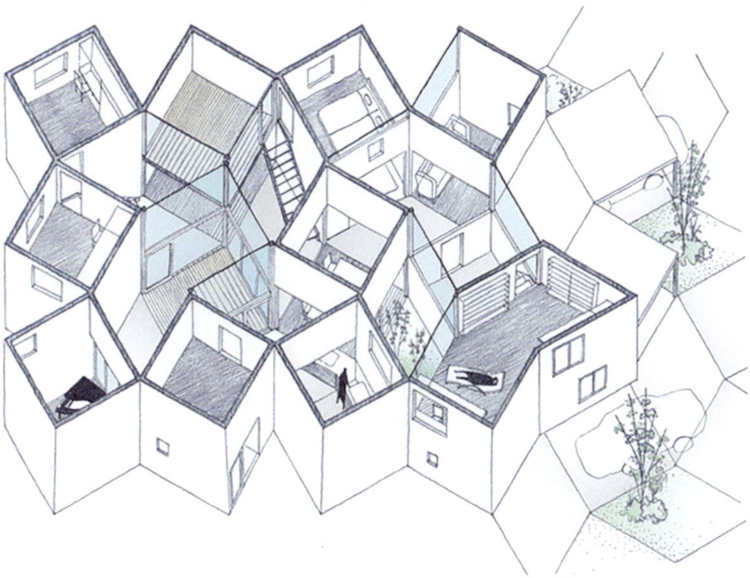

Fig. 4.8 A typical space design [6]

4.5 Benefits

Despite its growing popularity in architecture, generative design has its pros and cons. Generative design offers many advantages over traditional design methods. This technology enables the creation of thousands of design solutions in a short period, saving costs and expanding creative possibilities far beyond traditional methods. Generative design not only optimizes design and manufacturing methods but also saves production costs and enhances product value. Generative design helps designers reduce the time needed to create a product. Other benefits include the following [2, 4]:

- *Creative Designs*: Generative design is a creative process. It can create products with geometries and structures that even professional designers cannot imagine. This helps expand the boundaries of creativity and explore new possibilities in product design. As the design options produced by generative design tools are way beyond what the human mind can envisage, it helps one overcome creative barriers.
- *Speed*: Generative design also speeds up the design process. It allows the designer to work faster, make better-informed design decisions and provide them with data to justify specific design choices. What would take a human architect weeks to

Fig. 4.9. The Shanghai Tower in China [7]

achieve can be done in a matter of hours or even minutes with AI. This efficiency does saves time and leads to a better end product.

- *Efficiency*: Design algorithms enable architects to explore a vast design space efficiently, considering multiple factors such as performance, aesthetics, and user preferences simultaneously. The designs are created based on specific manufacturing constraints and requirements, ensuring feasibility and efficiency in production.
- *Reduced Costs*: By optimizing material usage and manufacturing, generative design helps reduce production costs. This not only reduces waste but also increases the economic value of the final product. Designs produced by generative modeling do not require additional simulation or testing. The designs adhere to safety norms and are ready for execution because simulation is built into it as a feature. Thus, an advantage of generative design is that it leads to lower costs and frees up budget for other processes.

- *Reduced Weight*: One of the major advantages of generative design is its ability to create lighter designs while maintaining high strength and performance. This is particularly important in industries such as aerospace and automotive, where reducing product weight can lead to fuel savings and improved overall performance.
- *Design Automation*: Design must be done by humans, and computers are there to automate tasks and save time for us. Generative design accelerates the design process by automating repetitive tasks and exploring a vast array of design alternatives. By defining design objectives and constraints, architects can leverage generative design algorithms to generate and evaluate thousands of design options quickly, facilitating informed decision-making and design optimization. It is not surprising that generative design, which is a step ahead of parametric modeling, is capable of automated adjustments.
- *Optimizing Performance*: Generative design enables architects to optimize building performance by evaluating design alternatives against performance criteria such as energy efficiency, structural integrity, and occupant comfort.
- *Customization*: Generative design facilitates the creation of customized and personalized architectural solutions that respond to specific site conditions, user needs, and project requirements. By incorporating user preferences and constraints into the design process, architects can create spaces that are tailored to the unique needs and aspirations of their clients.
- *Countless Choices*: One major advantage of using generative design as a design tool is that it can generate an infinite set of possible design solutions. A team can literally try hundreds of options to find the best solution. Backed by the storage capacity of cloud, an advantage of generative design is that the process produces hundreds of design iterations despite specified constraints. This is well beyond what an architect or designer can manually create. The advantage of generative design is that it helps create multiple creative options within a short period—a task which is humanly impossible.
- *Free Up Time*: An advantage of generative design is that it helps architects skip the need of coming up with designs and manually navigating the design options. It frees up their time so that they can focus on more human and practical engagements. It allows them to take up a more strategic role.
- *Human Factors*: By considering human factors such as ergonomics, circulation, and comfort, architects can design spaces that enhance the quality of life for occupants and foster a sense of connection and belonging.
- *Sustainable Design*: Generative design in sustainable design is an effective approach addressing energy efficiency and climate change at the early design stage. It integrates environmental principles with algorithms, enabling exploration of countless design alternatives to enhance energy performance, reduce carbon footprints, and minimize waste. This not only minimizes waste and costs but also improves efficiency, making designs sustainable.
- *Improved Collaboration*: With cloud computing capabilities, generative design architecture software enables better collaboration between team members, stakeholders, and even clients to have an integrated, streamlined design process in

place. While it is great that architects and other members of the team can view the designs on a shared platform, without a clear and transparent framework for decision-making and feedback, this can lead to confusion or frustration.

4.6 Challenges

Challenges such as data interoperability, computational complexity, and skill gaps pose obstacles to widespread adoption of generative design in practice. Addressing these challenges requires investment in training, infrastructure, and collaboration across disciplines to ensure successful implementation and integration into existing workflows. While embraced by a large part of the AEC community, generative design and its reliance on artificial intelligence has raised some concerns. Balancing multiple design objectives while limiting computational costs adds further challenges for designers. Other challenges include the following [1, 2, 8]:

- *Human Creativity*: While generative design offers unprecedented opportunities for automation, it is essential to maintain a balance between computational efficiency and human intuition to ensure the integrity. Architects must resist the temptation to rely solely on algorithmic solutions and instead embrace technology as a tool to augment and enhance human creativity and ingenuity.
- *Making Humans Redundant*: Architects wonder if the machine learning-enabled software feature would replace the need for designers and architects in the future. Generative design proponents believe such fears are unfounded. They believe that architects need to expand their skill sets. Generative design is just another tool in the toolbox. After all, there is a lot more to the architectural profession than constraint-based design.
- *Emerging Technologies*: Advancements in emerging technologies such as artificial intelligence, machine learning, and cloud computing are poised to further enhance the capabilities and accessibility of generative design tools. These technologies will enable architects to tackle increasingly complex design challenges and explore new frontiers in architectural expression and innovation.
- *Ethical Issues*: Generative design relies on AI, automation, and the use of potentially sensitive data. This can raise questions about ownership, responsibility, and liability. There could also be issues around privacy, security, and consent of the data sources.
- *Algorithm Bias*: Concerns regarding the loss of human agency and the potential for algorithmic bias necessitate careful consideration and ethical oversight in the development and implementation of generative design tools. Architects must remain vigilant and critical of the outputs generated by algorithms, recognizing their limitations and biases and actively engaging in the design process to ensure that human values and priorities are upheld.
- *Shortlisting Options*: Interestingly, a disadvantage of generative design is that it throws up too many options. There are software designed to filter options based on cost or other parameters. But there many software that do not provide an

advanced filtering option. Architects seldom have the time and money to explore all the possibilities and have to settle on a design at some point. They do not have the time or interest to tediously filtering out designs. Thus, sometimes, it can get a little too overwhelming.

- *Manual Intervention*: Though largely automated, a disadvantage of generative design is that it does require architects to be capable of leveraging machine learning and AI to their advantage. Otherwise, the resulting designs may not stand up to the high expectations we have, out of generative design.
- *Quality of Data*: As generative design tools rely on data from a real-world setting, this data needs to be as accurate as possible. Inaccurate, outdated or incorrect information can lead to flawed designs that do not meet the project requirements.
- *Skills Gaps*: Generative design tools are complex and sophisticated and architects can find this daunting if they have not used them before. This means that many would need to retrained, otherwise the industry could face an ongoing skills shortage. Generative design tools operate on sophisticated algorithms, making them technically complex and potentially daunting for professionals unfamiliar with computational tools. Architects need to invest their time and energy in learning and upskilling in the realms of coding and generative design tools.
- *Software Compatibility*: It can be tough to find the right generative design software that is compatible with existing systems and workflows, as this often requires specialized platforms. As such, architects will need to evaluate the compatibility of the software to make sure it will support their design goals.

4.7 Conclusion

Generative design is transforming the architecture industry and built environment. It has become a paradigm shift in architecture. It enables architects to create more efficient, innovative, and personalized designs in less time. It is revolutionizing the way architects approach creativity, efficiency, and problem-solving. Although generative design is still a relatively new design technology, it is slowly making its way into the world of architecture.

Generative design is making a name in the world of contemporary architecture. In order to take generative design to the next level in the AEC, the industry must first upgrade its technology and data literacy to understand the outcomes that result from using these powerful tools. The integration of generative design into architectural curricula and professional practice is reshaping how architects learn, design, and collaborate. More information on the integration of generative design into the architecture industry is available from the books in [9–12].

References

1. T. Kaarwan, BIM and generative design: exploring algorithmic design approaches. https://www.kaarwan.com/blog/architecture/BIM-and-generative-design?id=396
2. Advantages & disadvantages of generative design (2020). https://www.infurnia.com/blog/generative-design-pros-cons
3. M.N.O. Sadiku, P.A. Adekunte, J.O. Sadiku, Generative design in architecture. Int. J. Trend Sci. Res. Dev. **9**(1), 377–386 (2025)
4. T. Chu, Benefits of generative design in design and manufacturing. https://vinnotek.com/blogs/danh-muc-blog-tieng-anh/benefits-of-generative-design-in-design-and-manufacturing#:~:text=Generative%20design%20not%20only%20optimizes,that%20traditional%20design%20cannot%20achieve.
5. Generative design for architectural space planning. https://www.autodesk.com/autodesk-university/article/Generative-Design-Architectural-Space-Planning
6. E. Tovar, Generative space design: exploring 8 transformative tools in architecture (2023). https://www.archdaily.com/1001757/generative-space-design-exploring-8-transformative-tools-in-architecture
7. P. Sharma, How generative design is shaping the future of architecture? (2023). https://parametric-architecture.com/how-generative-design-is-shaping-the-future-of-architecture/?srsltid=AfmBOoodkHhwQw8rVG0mB8SY7cq8qGmQwIilrKL7qRh7NoHodH-47iBa
8. The architect's guide to generative design (2024). https://www.google.com/search?q=The+architect%E2%80%99s+guide+to+generative+design%2C+15+July%2C+2024%2C+p.116&rlz=1C1UEAD_enUS1082US1082&oq=The+architect%E2%80%99s+guide+to+generative+design%2C+15+July%2C+2024%2C+p.116&gs_lcrp=EgZjaHJvbWUyBggAEEUYOTIHCAEQIRiPAjIHCAIQIRiPAtIBCDMyMDBqMGo3qAIAsAIA&sourceid=chrome&ie=UTF-8
9. E. Lanzara, *Generative Design Strategies for Customizable Prototypes: Academic Research and Entrepreneurial Education* (IGI Global, 2021)
10. A. Agkathidis, *Generative Design: Form-finding Techniques in Architecture (Form + Technique)* (Laurence King Publishing, 2016)
11. C. Cogdell, *Toward a Living Architecture? Complexism and Biology in Generative Design* (University of Minnesota Press, 2019)
12. X. Marsault, *Eco-Generative Design for Early Stages of Architecture* (Wiley-ISTE, 2017)

Chapter 5
Robotics in Architecture

Robotics is not about making machines to serve us. It's about creating machines that can be our partners.—Cynthia Breazeal.

5.1 Introduction

Robotics has always inspired a keen sense of curiosity and interest among individuals.

An important element in the advancement of technology in recent years is the inclusion of robotics in our daily lives. The use of robotics in architecture and construction was once the stuff of science fiction. Robotics has been utilized now in several sectors of society. Robotics in architecture is an evolving technology implementing robotics for accurate models and structures. In the early days, robots were used for simple tasks like bricklaying. After that, people started using them to build big structures like bridges and buildings. Now, robots are being used on construction sites to create modern and elegant buildings. Robotics and artificial intelligence represent the next cutting edge technology that will transform the fields of architecture and design.

Robotics is the discipline of designing and constructing machines, called robots.

A robot is an autonomous mechanical device that is designed to sense its environment, carry out computations to make decisions, and perform actions like humans in the real world. It is a system that contains sensors, control systems, power supplies, and software, all working together to perform a task. Robotics is a relatively young field with highly ambitious goals. It is producing a huge range of devices, from autonomous vacuum cleaners to military drones [1]. Robots come in a wide range of shapes and sizes based on the role they are meant to fill on-site.

Robots today serve as mechanized interfaces between the digital and the physical worlds, turning data into action. They have played important roles in manufacturing for decades, but a new wave of robotics innovation focuses on uses in architecture and construction. Figure 5.1 shows a robot engineer at work [2].

Robots are programmable automated mechanisms that help out in the process of digital fabrication. Robotics refers to any machine or technology system that

M. N. O. Sadiku, *Emerging Technologies in Architecture*,
https://doi.org/10.1007/978-3-031-99509-5_5

Fig. 5.1 A robot engineer at work [2]

performs a set of tasks that would normally be done manually by humans. Robots are transforming the way architects approach their designs. They are already being used in virtually every step of the building design process, from initial site analysis to construction. The architecture and construction sector utilizes robotics in designing and producing precise designs and products. Advances in robotics can change how and what we build, and may eventually empower us to reinvent conventional construction processes entirely [3].

This informative chapter delves into the innovative ways robotics is being integrated into architecture. It begins with explaining a robot. It describes architectural robots and provides some of their applications. It highlights the benefits and challenges of robotics in architecture. The last section concludes with comments.

5.2 What is a Robot?

The word "robot" was coined by Czechriter Karel Čapek in his play in 1920. Isaac Asimov coined the term "robotics" in 1942 and came up with three rules to guide the behavior of robots and later added the zeroth law [4]:

- Law 0: A robot may not injure humanity or through inaction, allow humanity to come to harm.
- Law 1: Robots must never harm human beings,

- Law 2: Robots must follow instructions from humans without violating rule 1,
- Law3: Robots must protect themselves without violating the other rules.

Robots are becoming increasingly prevalent in almost every industry, from healthcare to manufacturing. Figure 5.2 indicates that robotics is one of the branches of artificial intelligence.

Although there are many types of robots designed for different environments and for different purposes/applications, they all share four basic similarities [5]:

(1) All robots have some form of mechanical construction designed to achieve a particular task;
(2) They have electrical components which power and control the machinery;
(3) All robots must be able to sense its surroundings; a robot may have light sensors (eyes), touch and pressure sensors (hands), chemical sensors (nose), hearing and sonar sensors (ears), etc.
(4) All robots contain some level of computer programming code.

Programs are the core essence of a robot since they provide intelligence. There are three different types of robotic programs: remote control, artificial intelligence, and hybrid. Some robots are programmed to faithfully carry out specific actions over and over again (repetitive actions) without variation and with a high degree of accuracy.

Robotics is an interdisciplinary field that involves the design, construction, operation, and use of robots. It is a branch of engineering and computer sciences that includes the design and use of machines that are capable of performing programmed tasks without human involvement. The field develops machines that can efficiently carry out various tasks, can automate tasks, and do various jobs that a human might not be able to do. Robots could someday be our drivers, companions, collaborators, teachers, specialists, and exploration pioneers [6].

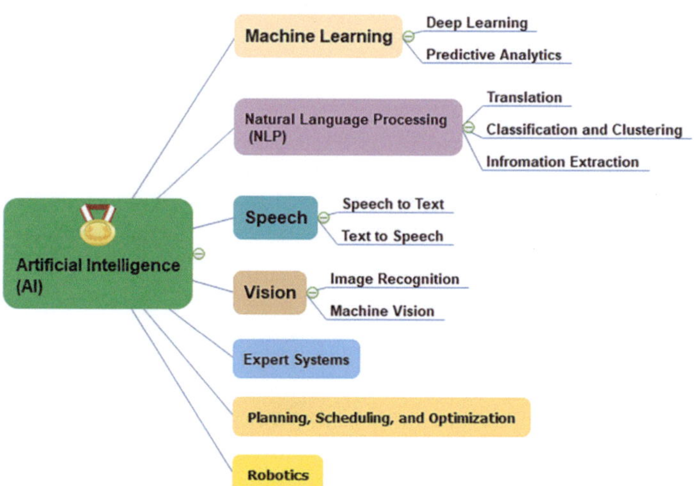

Fig. 5.2 Robotics is one of the branches of artificial intelligence

Fig. 5.3 A typical use of robots architecture [7]

5.3 Architectural Robots

From household assistants like Roomba to personal assistants like Siri and Alexa, to companion robots, these ingenious devices, empowered by artificial intelligence and automation, have found a place in almost every industry today. Robot refers to any machine or technology system that performs a set of tasks that would normally be done manually by humans. In the field of architecture, robotics has begun to revolutionize the way different types of buildings are designed, constructed, and even maintained. Robotics in architecture is the use of robots to design, construct, and research buildings, allowing architects to create more complex and innovative designs. It involves the integration and application of automated machines—or robots—into architectural design processes. A typical use of robots architecture is shown in Fig. 5.3 [7].

Robotics in architecture is a growing field. It uses machines to design, build, and research. These robots can bring new ideas to how we make buildings. Imagine a world where robots and architects collaborate. The integration of robotics in architectural processes allows for the completion of complex and precise tasks, optimizes resources, and helps explore alternative methods of design and construction.

5.4 Applications of Robotics in Architecture

Robotics in architecture and construction is transforming the way architects construct and design buildings. Architects are beginning to explore how this technology can be used for more than just basic modeling, construction, assembly, on-site troubleshooting, and energy efficiency. Robotic applications are already transforming the way architects construct and design buildings. Common areas of application of robots in architecture include the following [8]:

- *On-site Assembling*: Construction on-site is even faster with robots. Robots can assemble pieces on-site, replacing cranes in some cases. Long robotic arms replace cranes on some construction sites. Robotics engineers are also beginning to use drones and robotic helicopters to put small bricks and other pieces into place that would normally require the use of a hydraulic lift or elaborate scaffolding. They are implanting long robot arms in construction sites to lay bricks, arrange structural elements, and more. Figure 5.4 shows an industrial robotic arm for assembling [9].
- *Robotic Fabrication*: This is revolutionizing the field of architecture and design, transforming how we create and construct buildings, furniture, and other structures. The robotic fabrication method in architecture refers to the use of robotic systems and emerging technologies to design, construct, and assemble architectural structures. It involves the integration of advanced digital technologies, such as computational design processes and tools and robotic arms, into the architectural fabrication process. It has played a crucial role in creating innovative and awe-inspiring creations. It offers architects and designers greater flexibility, precision, and efficiency in realizing complex and innovative architectural forms. Robotic fabrication is revolutionizing the field of architecture and design by transforming

Fig. 5.4 An industrial robotic arm for assembling [9]

Fig. 5.5 The MX3D Bridge in Amsterdam [10]

how we create and construct buildings, furniture, and other structures. The MX3D Bridge in Amsterdam is a significant example of robotic fabrication in architecture. It is displayed in Fig. 5.5 [10]. Robotic fabrication optimized the design by distributing material strategically, reducing waste and weight while ensuring structural integrity. Robotic fabrication improved the construction efficiency of the bridge.

- *Prefabrication*: This is the case in which buildings are either assembled onsite or prefabricated and brought to the site whole, dramatically reducing or even eliminating on-site construction time. By enabling a more efficient work cycle, robots pose great potential for improving the speed of the prefabrication process. Prefabrication is crucial in disaster stricken areas or refugee camps, where on-site construction is detrimental/impossible and importing whole structures, parts, or even materials, can be complicated and expensive.
- *Robotic Automation*: This streamlines construction, increasing efficiency and reducing time spent on repetitive tasks. Automation is no longer a distant concept as robots collaborate with humans in the creation process, redefining roles and methods. Automated construction could become the norm. Robots with tools can make complex shapes with ease, which brings more creative freedom for architects. Robots can be programmed to perform precise and repetitive tasks such as cutting and shaping materials, manufacturing custom components, or constructing complex structures. Figure 5.6 displays robots working on a construction site [11].
- *Robot Science Museum*: The Robot Science Museum in Seoul, South Korea, will be built using robotic technology and drones. This is a powerful example of robotics in architecture. This project helps us see the future of building design and construction. Figure 5.7 shows the museum [12].

Fig. 5.6 Robots working on a construction site [11]

- *3D Modeling*: Robots can create accurate 3D models of buildings using 3D printing. Robots build small-scale models of buildings with great precision through the use of 3D printing. Architects can actually test different types of materials in these models to determine which materials will be strongest or most flexible.
- *Construction Parts*: Robots can mill and shape custom building parts. Robotic manufacturing tools mill both large and small custom building pieces with great accuracy and detail.
- *Sustainability*: Robots can be embedded in buildings to monitor temperature, lighting, air quality, and more. Architects can embed robots directly into the building itself that monitor temperature, lighting levels, air quality, motion, and more.

Fig. 5.7 Robot science museum [12]

- *Material Handling*: Robots can test materials in 3D models to determine their strength and flexibility. Robotic systems optimize material usage, minimizing waste and promoting sustainable construction practices.

5.5 Benefits

Robot application in architecture industry poses several advantages and benefits in comparison to human workers. Robots can often perform the same tasks much faster and can even work 24/7 on projects without getting fatigued. Robots can also play a significant role in the interaction and adaptation of buildings to their environment. They can perform dangerous construction tasks (like demolition or complex crane work), which lessens the risk of injury to human workers. Other benefits of robotics in architecture include the following [10]:

- *Precision*: Precision is at its best with robotics. Robots can perform tasks with great accuracy. The technology is less prone to error; the products robots build are more consistent in shape, construction, and overall quality. Robots can build buildings and construction parts accurately without any errors. Accuracy and precision brought by robots can help create designs and products of better quality. Robots utilized in designs create architecture with minimal issues or inconsistencies. Products and structures designed are consistent in terms of their shape, form, and quality. Complex timber structures that require precision and accurate construction are convenient now by utilizing robots. Complex forms and structures require precision and accuracy now they can be easily accomplished utilizing robots. A typical complex structure that requires a robot is shown in Fig. 5.8 [7].

- *Long Work Span*: Humans are prone to distractions, stamina, and attention; hence, the working span is reduced compared to robots. Robots can help reduce the time required for construction due to the long work span achieved by robots.
- *Speedy Construction*: Robots make a man's life easier for every individual, especially in the construction sector. The time taken to build a structure by a robot is less in comparison to manual labor.
- *Avoids Mishaps*: The probability of errors occurring is less; hence, the mishaps or accidents prone to occur are also less. Accidents that are prone to occur through manual labor are reduced and avoided considerably by robots.
- *Flexibility*: Robots can perform tasks that would be difficult or costly to do manually. Robotic systems are flexible and can be created to solve a wide range of architectural problems. The robotic systems can be programmed to perform various tasks, including cutting, milling, 3D printing, welding, and assembly, allowing for a wide range of construction possibilities.
- *Cost Reduction*: A robot is a more proficient process that could reduce the time and cost of construction. Robots save time, cut costs, and reduce waste when they build. Besides shaping and size, robots will help us save materials.
- *Collaboration*: When robots and humans work together, they can do more than either could do alone. They make it safer for workers on site because robots can take over risky jobs. Robots are learning from each other as well. Figure 5.9 shows collaboration between robots and humans [7].

Fig. 5.8 A typical complex architectural structure using a robot [7]

Fig. 5.9 Collaboration between robots and humans [7]

- *Novel Designs*: Machine learning is still very limited when it comes to creativity and spontaneous design inspiration. Robots could come up with fascinating solutions to complex problems that architects had never considered before. Robots can help architects explore new ways of designing and constructing buildings.
- *Customization*: The integration of robots in architecture offers new creative possibilities, greater resource efficiency, and the ability to tackle architectural challenges in innovative ways. Robotics allows for customization of the living spaces and reconfiguration of the structure according to individual preferences. This is typically shown in Fig. 5.10 [13].

5.6 Challenges

Using robots in architecture has some downsides. Climate change could alter the environment to such a great extent that habitable land becomes much more scarce. Construction sites present various safety hazards, including moving machinery, heavy equipment, and unpredictable working conditions. Other challenges include the following [14]:

Fig. 5.10 Robotics allows for customization [13]

- *Human Touch*: Human architects are still indispensable in the design process. The biggest limitation of robotics in architecture and construction is that it lacks a human touch, primarily when it comes to creativity. Architects have a rich experience and are able to use the breadth of their experiences to come up with new ideas.
- *Cost*: Cost is one big issue. Implementing robotic systems in construction often necessitates an upfront investment, which can be a barrier for small construction companies. This relates to the costs that construction companies must incur to adopt robotics in construction. Cost can be high and they take out the human touch from designs.
- *Resistance to Change*: The construction industry has traditionally relied on manual labor, which can cause resistance to adopting new technologies. Skepticism, fear of job displacement, and lack of awareness about the potential benefits of automation may hinder the acceptance and implementation of automated solutions.
- *Energy Efficiency*: With buildings accounting for nearly 40 percent of US energy consumption, new approaches to ensuring energy efficiency are in high demand.
- *Lack of Flexibility*: Construction projects often involve unique and customized tasks that require flexibility, as well as collaboration and teamwork among various trades and skilled workers. The ability to communicate, coordinate, and make adjustments on the fly is crucial for seamless project execution.

5.7 Conclusion

Robotics technology will soon become a fundamental part of the architectural process. Architects who embrace this dynamic technology now will be better equipped to design the most efficient buildings of the future. When architects embrace this technology, they can create buildings at a lower cost and in record time. However, robots should not inspire architecture for the sake of robots, they should simply assist in architecture for the sake of people.

The Association for Robots in Architecture was founded in 2010 with the goal of making robots accessible to the creative industry. The Association is a highly transdisciplinary network of robot users around the world [15]. Robotics have been part of engineering curricula in several institutions for many years, and now the invasion is encroaching on architecture majors as well.

Robots are changing architecture. They help design and build in new ways. Robotics is constantly evolving and innovative robotics breakthroughs will shape the architecture industry of the future. In the future, robotics and technology utilization in building architecture and construction will be an important part of the construction sector. More information on robotics in the architecture industry can be found in the books in [16–30] and the following related journals:

- *Robotica*
- *Robotics*
- *Robitics and Autonomous*
- *Robotics and Computer-Integrated Manufacturing,*
- *Advanced Robotics*
- *Autonomous Robots*
- *Automation in Construction*
- *Journal of Robotics*
- *Journal of Robotic Systems*
- *Journal of Robotic Surgery*
- *Journal of Robotics and Mechatronics*
- *Journal of Intelligent & Robotic Systems*
- *Journal of Mechanisms and Robotics-Transactions of the ASME*
- *Journal of Automation, Mobile Robotics and Intelligent Systems*
- *Journal of Future Robot Life*
- *IEEE Robotics and Automation Letters*
- *IEEE Transactions on Robotics*
- *International Journal of Medical Robotics and Computer Assisted Surgery*
- *International Journal of Robotics Research*
- *International Journal of Social Robotics*
- *International Journal of Humanoid Robotics*
- *International Journal of Advanced Robotic Systems*
- *Science Robotics*
- *Soft Robotics*

References

1. S. Brezgov, Robots in education: is the educational revolution just around the corner? (2020). https://scholarlyoa.com/robots-in-education-is-the-educational-revolution-just-around-the-corner/
2. The top 10 careers in robotics in 2022 (2022). https://www.neit.edu/blog/careers-in-robotics
3. M.N.O. Sadiku, O.D. Olaleye, J.O. Sadiku, Robotics in architecture. Int J Trend Res Dev **12**(1), 89–94 (2025)
4. Human–robot interaction, *Wikipedia,* the free encyclopedia. https://en.wikipedia.org/wiki/Human-robot_interaction
5. Robotics, *Wikipedia*, the free encyclopedia. https://en.wikipedia.org/wiki/Robotics
6. M.N.O. Sadiku, U.C. Chukwu. A. Ajayi-Majebi, S.M. Musa, A primer on robotics. Int. J. Trend Sci. Res. Dev. **6**(7), 614–621 (2022)
7. S. Arpitha, Robots in architecture: transforming architecture with precision and efficiency (2023). https://parametric-architecture.com/robots-in-architecture-transforming-architecture-with-precision-and-efficiency/?srsltid=AfmBOor7zvD5VC_wEXztQr-yCbbHp0rOj2t8mP_zw0kz3mn40lkuVTPD#google_vignette
8. Robotics in architecture and construction: An industry shift (2019). https://hmcarchitects.com/news/robotics-in-architecture-and-construction-an-industry-shift-2019-10-23/
9. M. Waibel, Architects using robots to build beautiful structures (2011). https://spectrum.ieee.org/robots-in-architecture
10. J. Goyal, 5 Innovative examples of robotic fabrication in modern architecture (2025) (2023). https://www.novatr.com/blog/robotic-fabrication-in-modern-architecture
11. S. Lucas, 10 AI applications in architecture and design (2024). https://futurearchi.blog/en/ai-applications-architecture-design/
12. Architectural robotics: The latest architecture and news (2022). https://www.archdaily.com/tag/architectural-robotics
13. Applications of robotics in the field of architecture. https://pertanto.com/en/applications-of-robotics-in-the-field-of-architecture/#:~:text=Robots%20can%20be%20programmed%20to,or%20costly%20to%20perform%20manually.
14. What are the barriers to adopting robotics in construction (2023). https://www.dustyrobotics.com/articles/what-are-the-barriers-to-adopting-robotics-in-construction
15. Creative robotics in architecture. https://www.autodesk.com/autodesk-university/article/Creative-Robotics-in-Architecture
16. M.N.O. Sadiku, *Robotics and Its Applications* (Lambert Academic Publishing, Moldova, Europe, 2023)
17. K.E. Green, *Architectural Robotics: Ecosystems of Bits, Bytes, and Biology* (The MIT Press, 2016)
18. A. Figliola, A. Battisti, *Post-Industrial Robotics: Exploring Informed Architecture* (Springer, 2020)
19. D. Reinhardt, R. Saunders, J. Burry (eds.), *Robotic Fabrication in Architecture, Art and Design 2016* (Springer, 2016)
20. G. Retsin et al., *Robotic Building: Architecture in the Age of Automation* (Detail, 2019)
21. J. Flux, *End-to-End Differentiable Architecture: Structuring Deep Reinforcement Learning for Robotics Control (Mastering Machine Learning)* (Independently Published, 2024)
22. F. Blanciak, *Tokyoids: The Robotic Face of Architecture* (The MIT Press, 2022)
23. K. Oosterhuis, *Ia #5-Robotics in Architecture* (Jap Sam Books, 2012)
24. J. Braumann, S. Brell-Çokcan, *Rob: Robotic Fabrication in Architecture, Art, and Design* (Springer, 2013)
25. A.J. Wit, M. Daas (eds.), *Towards a Robotic Architecture* (ORO Editions, 2018)
26. J. Willmann et al. (eds.), *Robotic Fabrication in Architecture, Art and Design 2018* (Springer, 2018)
27. F. Gramazio, M. Kohler, J. Willmann, *The Robotic Touch: How Robots Change Architecture* (Park Books, 2014)

28. H. Jebelli et al. (eds.), *Automation and Robotics in the Architecture, Engineering, and Construction Industry* (Springer, 2022)
29. F. Gramazio, M. Kohler (eds.), *Made by Robots: Challenging Architecture at a Larger Scale* (Wiley, 2014)
30. A. Bijl et al. (eds.), *CAD and Robotics in Architecture and Construction: Proceedings of the Joint International Conference at Marseilles, 25–27 June 1986* (Springer, 1986)

Chapter 6
Drones in Architecture

When everything seems to be against you, remember that an airplane takes off against the wind, not with it.—Anonymous.

6.1 Introduction

As far as architecture is concerned, there is always a change in technology and design. There is no denying the technological transformation that architecture is currently undergoing. From design software to electronic utilities and applications, architecture is growing into one of the most technically savvy professions in the world. One of the notable technologies that has accelerated the transformation is the drone.

Unmanned aerial vehicles (UAVs), commonly called drones, are gaining in popularity among professionals working in the Architecture, Engineering, and Construction (AEC) industry, which an industry that involves the design, engineering, and construction of buildings, infrastructure, and other physical structures. Drones were initially designed and built for military purposes, but they have been lately designed to serve non-military fields. Drones made their first appearance to the general public in 2006. Some years ago, using commercial drones would have been laughably unrealistic, but now they are in daily use on construction sites around the world. Drone technology is revolutionizing the design, construction, and maintenance of buildings. Today, large AEC companies like AECOM are regularly using drones to survey sites, monitor resources, and track project progress. International companies like Kespry and senseFly offer drones that are relatively easy to use and come with full-support packages [1].

Architecture is the art and science of designing buildings and other physical structures. Architects give importance to precision, and with drones architects no longer need to undergo rigid processes to get an aerial view of an area. Drones are already shaping the face of our cities, used for building planning, heritage, construction, and safety enhancement. Drone technology is playing a key role in redefining how we maintain, build, and design properties. Architects mainly use drones for generating

M. N. O. Sadiku, *Emerging Technologies in Architecture*,
https://doi.org/10.1007/978-3-031-99509-5_6

photographic views and developing their projects from particular angles. Researchers and architecture companies are already experimenting with building with drones [2].

This chapter focuses on drone applications in the architecture industry. It begins with explaining a drone. It describes drones in architecture and provides some of their applications. It highlights the benefits and challenges of drones in architecture. The last section concludes with comments.

6.2 What is a Drone?

At least three terms are used to describe drones, depending on how they are operated. The terms include Unmanned Aerial Vehicles (UAVs), Unpiloted Aircraft System (UAS), and Remote Piloted Aircraft System (RPAS). The FAA defines drones, also known as unmanned aerial vehicles (UAVs), as any aircraft system without a flight crew onboard. Drones include flying, floating, and other devices, including unmanned aerial vehicles (UAVs), that can fly independently along set routes using an onboard computer or follow commands transmitted remotely by a pilot on the ground. A typical drone is shown in Fig. 6.1 [3]. A drone is usually controlled remotely by a human pilot on the ground, as typically shown in Fig. 6.2 [3]. Drones can range in size from large military drones to smaller drones. Drones, previously used for military purposes, have started to be used for civilian purposes since the 2000s. Since then, drones have continued to be used in intelligence, aerial surveillance, search and rescue, reconnaissance, and offensive missions as part of the military Internet of things (IoT). Today, drones are used for different purposes such as aerial photography, surveillance, agriculture, entertainment, healthcare, transportation, law enforcement, etc.

Drones work much like other modes of air transportation, such as helicopters and airplanes. When the engine is turned on, it starts up, and the propellers rotate to enable flight. The motors spin the propellers and the propellers push against the air molecules downward, which pulls the drone upwards. Once the drone is flying, it is able to move forward, back, left, and right by spinning each of the propellers at a different speed. Then, the pilot uses the remote control to direct its flight from the ground [4].

Drone laws exist to ensure a high level of safety in the skies, especially near sensitive areas like airports. They also aim to address privacy concerns that arise when camera drones fly in residential areas. These include the requirement to keep your drone within sight at all times when airborne. In the United States, drones weighing less than 250 g are exempt from registration with civil aviation authorities. If your drone exceeds 250 g in weight, you will also require a Flyer ID, which requires passing a test [5]. It is necessary to register as an operator, be trained as a pilot, and have civil liability insurance, in addition to complying with various flight regulations, and those of the places where their use is permitted.

Most drones have a limited payload, usually under 11 pounds. Drones are classified according to their size. Here are the different drone types:

Fig. 6.1 A typical drone [3]

Fig. 6.2 A drone is usually controlled by operators on the ground [3]

- Nano Drone: 80–100 mm
- Micro Drone: 100–150 mm
- Small Drone: 150–250 mm
- Medium Drone: 250–400 mm
- Large Drone: 400 + mm

One of the emerging trends in drone use for factories is the utilization of LiDAR technology. LiDAR stands for Light Detection and Ranging. This technology provides accurate depth information essential for understanding the three-dimensional structure of the environment. LiDAR sensors emit laser beams to measure distances to objects, creating high-resolution 3D maps of the surrounding terrain and objects. The ability to capture detailed data through LiDAR technology has opened up opportunities for better predictive maintenance, reduction in inspection times, and overall cost savings [6].

6.3 Drones in Architecture

The Architecture, Engineering, and Construction (AEC) services sector provides a host of services surrounding the planning and building of commercial and residential projects that employ design and construction professionals in the planning and implementation of projects. One of the most compelling reasons to use a drone in architecture is aerial photography. Drones in architecture are redefining site analysis, inspections, surveying, and project documentation for architects. Properties located in scenic environments, such as those with mountain views, lakes, or extensive gardens, can significantly benefit from aerial shots.

Drones for architects have improved the architecture and construction industry, considering that it has given them the eyes that work best for jobs where precision and monitoring are essential. A drone will enable one to capture the tiniest of details and showcase a particular structure from every possible angle. The use of drone technology is becoming more and more common among architectural firms as a tool for site analysis, visualization, and project management. The advancement in drone technology is bound to influence architecture more than ever before. A typical example of drones in architecture is shown in Fig. 6.3 [7].

6.4 Applications of Drones in Architecture

Drones offer various functions throughout the building's lifecycle, including assessing site conditions, aiding in site planning, capturing construction progress photos, performing 3D scans of existing structures, creating virtual building models, assisting with maintenance, and supporting marketing efforts. The areas of drone applications include surveying and mapping, progress monitoring, safety monitoring, and asset maintenance. Common areas of applications include [8]:

- *Photography*: One of the best characteristics regarding the use of drones is their ability to capture hard-to-take images. It is impossible to grasp a land structure without seeing it in an aerial view. Before drones became helpful in architecture, construction companies would hire people to do an aerial view. Today, everyone

Fig. 6.3 A typical example of drones in architecture [7]

is using drones to photograph buildings from the sky. Drone photography showcases three things: unique perspective, location, and scale. Drones equipped with high-resolution cameras and sensors can capture detailed aerial images and data of the project site. Drone architecture photography produces precise images from an aerial perspective. Aerial photography captured by drones has become an asset to architects' building projects because it provides architects with a unique perspective of the site and its surroundings. Modern architecture often times looks stunning from the top down. Certain properties, particularly those with expansive lots, may require aerial photography to capture the entire space effectively. An aerial view of a building is shown in Fig. 6.4 [9].

- *Planning*: In site planning and layout, drones' aerial images combined with site plans provide accurate and visual representations of the building's features and their placement on the site. This helps designers and owners visualize the project and accommodate geographic features and variables accurately. With a more accurate photo of the actual land area, it is much easier to create a layout and design of the construction area. Drones in architecture enhance design visualization and allow design to become immersive and compelling.
- *Progress Monitoring*: It is very hard to keep track of construction projects as they involve many teams and can be spread out over a large site. Fixed cameras onsite may not be able to cover the entire site in detail, failing to capture critical progress information. Drones serve as vigilant overseers, monitoring construction progress and ensuring adherence to designs. They enable architects to monitor the

Fig. 6.4 An aerial view of a building [9]

construction process and track progress. They can create accurate 2D maps and 3D models of the site on-demand.

- *Building Inspection*: Traditional inspections often require teams to implement manual, rope-access techniques which require walking across rooftops and scaffolding. This is time-consuming and dangerous. It is often difficult to visualize the building in its entirety, and manual inspections do not produce standardized digital results. Drones can take high-resolution close-up photos of buildings to create detailed models for inspection. They can assist in building inspections, especially for large or complex structures. They can be used to inspect building facades, roofs, and other inaccessible areas without the need for manual intervention. Figure 6.5 shows the essential civic building of the future, a project designed by a student in Spain [10]. It shows how architecture can be "an engine of development and innovation.
- *Site Inspection*: The first and foremost consideration in architectural design is the physical context of any construction site. Site analysis and inspections are traditionally characterized by manual surveys, satellite images, and site visits, which take a lot of time and resources. Drones capture bird-eye views, helping architects understand topography, neighboring structures, and even shadow patterns. The use of drones has changed this working environment by providing a more accurate way of obtaining more efficient and accurate data. Drones fitted with high-definition cameras and laser imaging technology, often termed LiDAR (light detection and ranging), serve architects by creating detailed 3D maps and models of the site.

Fig. 6.5 The essential civic building of the future [10]

- *Drone Delivery*: The arrival of drone delivery services will see significant changes to buildings in our communities. A proposal for a drone delivery center is a serious example of an urban opportunity. Current models are still too weak to carry heavy materials, but they can easily carry very powerful cameras. A typical delivery drone is shown in Fig. 6.6 [11].

Fig. 6.6 A typical delivery drone [11]

6.5 Benefits

When it comes to architecture, the deployment of drones brings several benefits, including improved efficiency and reduced costs. Drones can survey sites, inspect construction quality, and photograph completed buildings, with more powerful capabilities in sight. They present significant opportunities to improve quality, reduce costs, improve decision-making, and mitigate safety risks in construction projects. It is a lot easier to fly a drone over a miles-long mining site that spans mountains and ravines than it is to send in a team of surveyors. Other benefits include the following [12]:

- *Reputation*: One of the foremost advantages is the improvement of design. Using drones to create high-quality 360° videos for your 3D modeling will also give you a competitive edge. Drones in architecture influence a company's reputation as they enhance their services. They give the impression that a company aims for a modernized workplace and services. You will truly impress your clients and turn them into repeat customers.
- *Risk Reduction*: Drones reduce the need to place staff and resources in a position of risk, allow repeatability to surveys, examinations and assessments, and offer a rapid and economical way of acquiring data.
- *Efficiency*: Drones bring efficiency, accuracy, and novel perspectives to the architectural process. Drones make architectural work more efficient as they lessen human effort by letting them control a drone remotely. Drones for architecture monitoring are a cost-effective solution to rising construction expenses. All the help that drones have offered in the architecture and construction industry saves many companies time and money. Embracing this technology not only spells efficiency and precision but sets the stage for futuristic, sustainable, and safer building designs.
- *Sustainable Architecture*: 21st-century architectural designs aim to contribute to sustainable architecture, which aims to help in the environmental movement. Drone architecture monitoring is made possible and lets architects create pleasing and environmentally friendly designs.
- *Cost Reduction*: Using drones is not just cost-effective but also saves time. It saves a lot of money and prevents any unwanted accidents that might cost lives.

6.6 Challenges

Despite the numerous advantages offered by drones, they have not gained widespread acceptance in facade inspection due to some restrictive local ordinances. The most important thing about flying a drone is to obtain a license or permit. If you fly your drone in a restricted area, you will get a fine. Other challenges include the following [12]:

- *Limitations*: Regulatory limitations include not being able to fly after dark, out of operator's line of sight, above certain altitudes, and near other structures or above crowds. Current models are still too weak to carry heavy materials, but they can easily carry very powerful cameras. If you are operating an unmanned aircraft that weighs less than 55 pounds, generally you may apply for a Part 107 waiver (special permission) to conduct your operation.
- *Regulations*: Some EU countries, including Czech Republic, France, Germany, Ireland, Italy, the Netherlands, Norway, Romania, and Sweden, have national regulations, while many others have still not addressed the issue.

6.7 Conclusion

The adoption of commercial drones in the AEC industry has been rapid partly because it aids in effectively sharing data and resolving challenges faced on construction sites. It might take a while before we see programmed drones build entire habitable buildings, but they can already contribute a great deal in surveying building sites, gathering site data, monitoring construction work, and creating marketing materials. Because drones can be used extensively in many phases of a project, they are becoming essential tools in the field of architecture.

Having a drone at the disposal of an architectural practice allows architects in the company to be proactive. Some architects dream of using drones to actually build a building. Given the rate at which drone technology has advanced in the past decade and the amount of money being invested in the area, it is likely that the dream will become reality. More information about drones in the architecture industry can be found in the books in [13–15].

References

1. M.N.O. Sadiku, P.A. Adekunte, J.O. Sadiku, A primer on drones. Int. J. Trend Sci. Res. Dev. **8**(4), 212–220 (2024)
2. M.N.O. Sadiku, P.A. Adekunte, J.O. Sadiku, Drones in architecture. Int. J. Trend Sci. Res. Dev. **9**(1), 282–288 (2025)
3. Drone in construction & infrastructure. https://www.jouav.com/industry/drone-in-construction
4. How drones work and how to fly them (2024). https://dronelaunchacademy.com/resources/how-do-drones-work/
5. What are the main applications of drones? (2024). https://www.jouav.com/blog/applications-of-drones.html
6. Drones in manufacturing: A game-changer for industry. https://viper-drones.com/industries/infrastructure-drone-use/manufacturing/#:~:text=The%20integration%20of%20drones%20into,on%20manufacturing%20is%20no%20exception
7. Architecture and drones. https://www.re-thinkingthefuture.com/product-design/a6106-architecture-and-drones/#google_vignette

8. Architecture, engineering, and construction services. https://www.trade.gov/architecture-and-engineering-services#:~:text=The%20Architecture%2C%20Engineering%2C%20and%20Construction,multi%2Dbillion%2Ddollar%20projects.
9. When to use a drone for architectural photography: A guide for architects, designers, and photographers. https://www.austinlaruephotography.com/architecturalphotographyblog/when-to-use-a-drone-for-architectural-photography-a-guide-for-architects-designers-and-photographers#:~:text=One%20of%20the%20most%20compelling,significantly%20benefit%20from%20aerial%20shots.
10. S. Lam, This student project shows how droneports could be the train stations of the 21st century (2016). https://www.archdaily.com/801464/this-student-project-shows-how-droneports-could-be-the-train-stations-of-the-21st-century
11. Five ways drones will change the way buildings are designed (2022). https://theconversation.com/five-ways-drones-will-change-the-way-buildings-are-designed-195611
12. L. Grozdaic, How drones can be used in architecture (and how to use them without breaking the law) (2018). https://www.archdaily.com/886743/how-drones-can-be-used-in-architecture-and-how-to-use-them-without-breaking-the-law
13. M.N.O. Sadiku, *Drones and Their Applications* (Xlibris Corp., Bloomington, IN, 2025)
14. D. Tal, J. Altschuld, *Drone Technology in Architecture, Engineering and Construction: A Strategic Guide to Unmanned Aerial Vehicle Operation and Implementation*(Wiley, 2021)
15. P. Cureton, *Drone Futures: UAS in Landscape and Urban Design* (Routledge, 2020)

Chapter 7
Internet of Things in Architecture

Powering billions of connected devices, the Internet of things has significantly transformed the model of interaction between humans and technology. Being a great vehicle of change, IoT is also anticipated to enhance our daily lives in the upcoming days.—Anonymous.

7.1 Introduction

IoT connects sensor-based objects to a shared infrastructure. This infrastructure can be an extranet, the Internet, or a network that eliminates the need for a human-to-computer interaction to exchange data. The purpose of architecture is to serve the people and make day-to-day life more manageable. We can currently link almost every device to the Internet to provide new ways of easing our lives. IoT devices monitor everything from manufacturing equipment status to trends on social media. Your smartphone, your smartwatch, and your digital thermostat are all part of the Internet of things, meaning that they are all devices that connect to the Internet, and through the Internet, connect to each other and enable mutual communication. These devices connect to a larger computer network to transfer critical information.

The Internet of things has transformed how we interact with the world and how we use and retrieve information. For many people, the most relatable examples of IoT devices are items in the home like home voice controllers, doorbell cameras, and smart TVs. For example, the electric meters that allow professionals to read your meter remotely are IoT devices. In the construction and architecture fields, IoT devices in cameras and drones can monitor worksites, generate advanced digital renderings of building projects, and more [1].

The Internet of things (IoT) is a term that describes a wide variety of devices that connect to computer networks such as the Internet. IoT consists of network-enabled devices embedded with sensors and connected to the Internet. It is about turning physical objects into smart objects. It connects sensor-based objects to a shared infrastructure. It is expanding rapidly. IoT devices and systems are everywhere, from

M. N. O. Sadiku, *Emerging Technologies in Architecture*,
https://doi.org/10.1007/978-3-031-99509-5_7

the smartphones we carry with us, to the devices managing our home's temperature or lights. IoT devices and connectivity are impacting the architecture industry in several ways [2].

This chapter covers IoT applications in the architecture industry. It begins with providing an overview on IoT and its subset, industrial IoT. It describes some applications of IoT in architecture. It highlights the benefits and challenges of IoT in architecture. The last section concludes with comments.

7.2 Overview on Internet of Things

The concept of the Internet of things (IoT) has been around since the late 1990s, but it gained momentum in the 2000s with the rise of Internet-connected devices. The Internet began with some military computers in the Pentagon called Arpanet in 1969. It expanded throughout the 1980s as a set of four parallel military networks, each at a different security level. The core technology which gives the Internet its particular characteristics is called Transmission Control Protocol/Internet Protocol (TCP/IP), which is essentially a set of rules for communication [3].

Internet of Things (IoT) is a worldwide network that connects devices to the Internet and to each other using wireless technology. IoT is expanding rapidly and it has been estimated that 50 billion devices will be connected to the Internet by 2020. These include smart phones, tablets, desktop computers, autonomous vehicles, refrigerators, toasters, thermostats, cameras, alarm systems, home appliances, insulin pumps, industrial machines, intelligent wheelchairs, wireless sensors, mobile robots, etc. Figure 7.1 illustrates various applications of the Internet of things [4].

There are four main technologies that enable IoT [5]: (1) Radio-frequency identification (RFID) and near-field communication, (2) Optical tags and quick response codes: This is used for low cost tagging, (3) Bluetooth low energy (BLE), (4) Wireless sensor network: They are usually connected as wireless sensor networks to monitor physical properties in specific environments. Communications technologies in Internet of things are portrayed in Fig. 7.2 [6].

IoT technology enables people and objects to interact with each other. It is employed in many areas such as smart transportation, smart cities, smart energy, emergency services, healthcare, data security, industrial control, logistics, retails, structural health, traffic congestion, manufacturing, and waste management. The Internet of things is extensively developed world-wide with a focus on civilian applications such as electric power distribution, intelligent transportation, healthcare, industrial control, precision agriculture, environmental monitoring, etc.

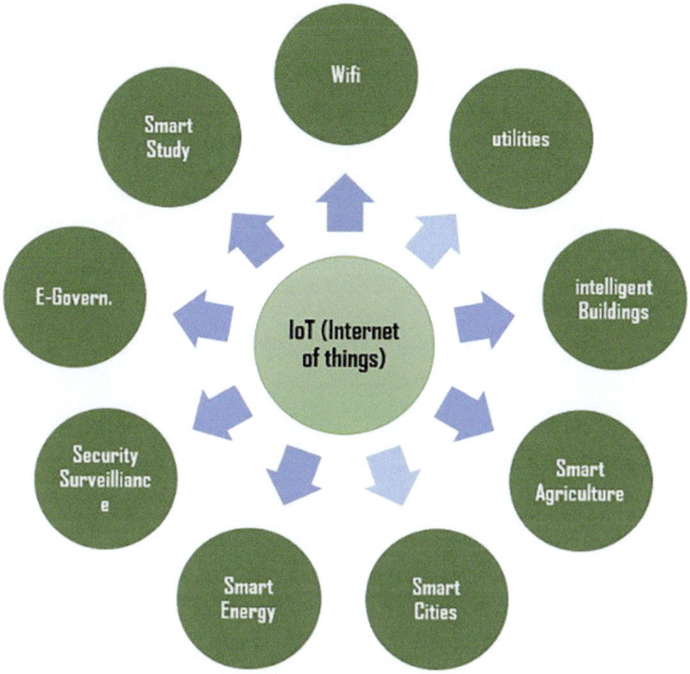

Fig. 7.1 Applications of the Internet of things [5]

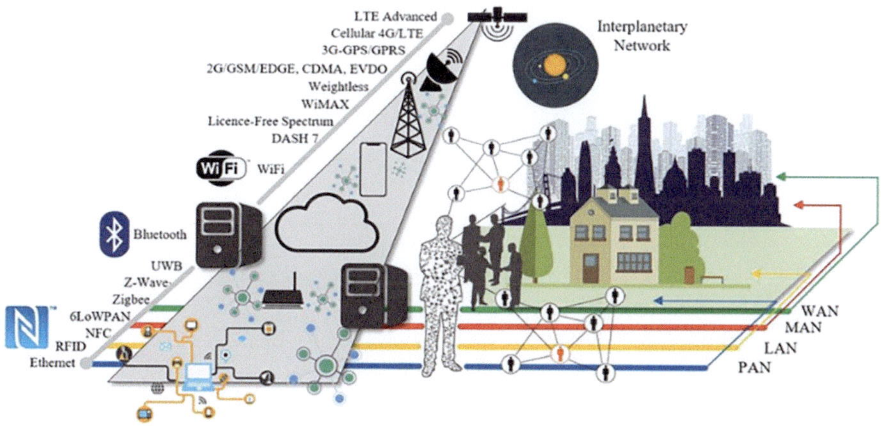

Fig. 7.2 Communications technologies in Internet of things [6]

7.3 Industrial Internet of Things

The growth of the internet of things (IoT) is drastically making impact on home and industry. While the IoT affects among others transportation, healthcare, or smart homes, the Industrial Internet of Things (IIoT) refers in particular to industrial environments. IIoT is a new industrial ecosystem that combines intelligent and autonomous machines, advanced predictive analytics, and machine-human collaboration to improve productivity, efficiency and reliability. It is bringing about a world where smart, connected embedded systems and products operate as part of larger systems [7].

The industrial Internet of things (IIoT) refers to the application of the Internet of things (IoT) across several industries such as manufacturing, logistics, oil and gas, transportation, energy/utilities, chemical, aviation and other industrial sectors. A typical industrial Internet of things is shown in Fig. 7.3 [8].

IIoT is often used in the context of Industry 4.0, the Industrial Internet and related initiatives across the globe. Industry 4.0 describes a new industrial revolution with a focus on automation, innovation, data, cyber-physical systems, processes, and people [9]. With Industry 4.0, the fourth industrial revolution is set on merging automation and information domains into the industrial Internet of things, services, and people. The communication infrastructure of Industry 4.0 allows devices to be accessible in barrier-free manner in the industrial Internet of things, without sacrificing the integrity of safety and security [10]. Figure 7.4 shows a typical representation of IoT [11].

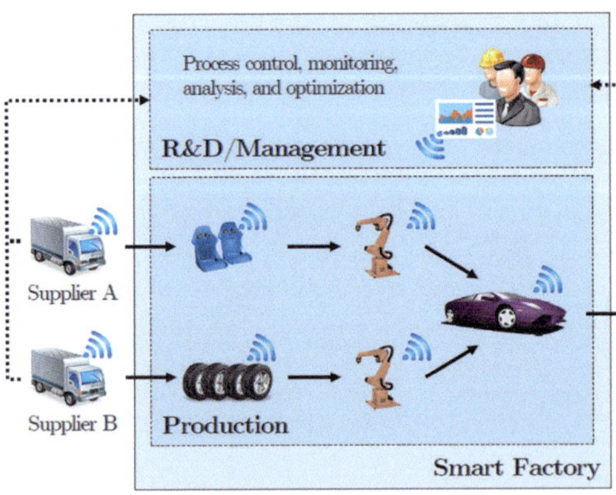

Fig. 7.3 A typical industrial Internet of things [8]

Fig. 7.4 A typical representation of IoT [11]

7.4 Applications of IOT in Architecture

Internet of things (IoT) is a reference to a network of connected devices and objects which have the capacity to gather information and share that data over the Internet. Like everything with the Internet of things, the use options are truly endless. Here are a few common ways the IoT in architecture and design is already making our lives easier [12, 13]:

- *Smart Buildings*: Due to their ability to control their own environment, smart buildings depend on a variety of sensors to regulate everything from temperature to lighting and security. Buildings are moving from passive assets to active built environments by becoming "smart." Smart or intelligent buildings are the most promising field of application for IoT. They provide much control over the environment and functions. They lead to the creation of highly efficient, high-capacity collaborative workspaces, where a smart system centralizes data from separate devices into a single one. Intelligent buildings can interact, learn and adapt by connecting people with technology and the environment. They have lent support to energy conservation and environmental goals as well. As an exemplary smart building, Salesforce Tower in San Francisco, is shown in Fig. 7.5 [12]. The building considers sustainability the main design feature and allows tenant employees to work wherever in the building. A smart building may use a IoT-powered rooftop solar panels that can support all of the building's power needs, as typically shown in Fig. 7.6 [1].
- *Smart Contracts*: A smart contract for a smart building is a digital agreement that automatically executes based on predetermined operations in your facility. It guides all the stakeholders involved in your building functions. By partnering with

Fig. 7.5 A typical smart building (Salesforce Tower in San Francisco) [12]

Fig. 7.6 A smart building with IoT-powered rooftop solar panels [1]

contractors or vendors through smart contracts, you can invest in smart systems and devices for various building functions.

- *Smart Homes*: A home is considered smart when all of the home's technical components are merged into a unified, intelligent system. The smart home is a concept to manage and simplify the usage of all technical equipment to create comfort, safety, and pleasure in one's own home. IoT in architecture enables smart homes to use a centralized system to control many devices. Smart homes contain a single automated system, linking all the devices such as a smartphone, tablet, laptop, door locks, cameras, lights, televisions, thermostats, house monitors, and even appliances like refrigerators as parts of the IoT technology. IoT plays a critical role in maximizing home security, remote controlling the home functions, and home management insights, making it a necessity rather than a facility. One of the main benefits of smart homes is that it facilitates the elderly living alone. A typical smart home is illustrated in Fig. 7.7 [12].
- *Smart Cars*: These refer to cars that can programmatically regulate speed and steering and rely on the human driver to take over in abnormal situations, such as when bad weather interferes with the car's sensors. Smart car ambition is directly driven by IoT, which will enable cars to become integrated carriers of digital transformation as more sensors, processors, and software are added to them. Currently, designers are searching for smart environmentally-friendly cars with low running costs. An example of a smart car is shown in Fig. 7.8 [12].

Fig. 7.7 A typical smart home [12]

Fig. 7.8 An example of a smart car [12]

- *Construction*: Traditionally, construction sites were isolated from offices, companies, and logistics chains. As a result, much time is wasted on every project due to unexpected delays and miscommunication. IoT is a powerful technology for changing industries, automating processes, and increasing return on investment. IoT technology has a lot to offer to the construction industry as it is a ready-to-change industry due to its challenges. IoT automates tasks that waste human hours. It functions in diverse construction fields, mostly with project progress tracking, like tracking material deliveries to the construction site or using wearable technologies to track employee movements to ensure they work efficiently. Figure 7.9 shows a typical construction site [12].
- *Site Monitoring*: When it comes to the Internet of things on the building site, think about sensors. IoT devices that monitor construction site conditions, automate tasks, and track deliveries connect these work sites to vital information. The devices attached and monitored with the IoT software and techniques can monitor calculations of humidity, temperature, and pressure to alert management about any potential happening that could cause damage and that needs immediate attention. Not only can IoT devices monitor your structures, they can also help keep your personnel safe.
- *Remote Monitoring*: IoT enables remote monitoring and notification by leveraging security systems like CCTV, access controls, and remote monitoring technologies. Ultimately, it helps promote building security and occupant safety. For advanced surveillance, one needs to equip smart building with networked cameras and sensors through IoT.
- *Equipment Tracking*: Small sensors can be easily connected to apps via the Internet of things, meaning that you can track and monitor a wide array of things. Affixing an IoT sensor to your tools makes them traceable in case of loss or theft. For example, if you work with prefabricated building sections, you can easily use the IoT to know the precise location of every piece.

Fig. 7.9 A typical construction site [12]

- *Building Information Modeling*: 3D building information modeling (or BIM) have become increasingly popular, as part of IoT in architecture and design, and are increasingly useful for engineers and builders. The integration of BIM and the Internet of things shows a great deal of promise.
- *Motion Detection*: Sensors that detect motion and vibration in large-scale structures can be used to help recognize dangerous situations and natural disasters before injury and loss of life occur. For example, specific disturbances that lead to catastrophic failures in bridges, dams, and buildings could be recognized before they occur.

7.5 Benefits

IoT is impacting architecture, construction, and engineering in a major way. IoT applications are making life simpler and more efficient. IoT a vital tool for the future for those working in the green building space. IoT ensures data centralization, regulatory compliance, and sustainability by consolidating information from various systems into a single platform. Other benefits include [14]:

- *Reliability and Resilience*: Consider the criticality of the IoT systems and data involved in the potential IoT solution. What level of downtime, if any, is an acceptable risk? What level of redundancy, failover, and disaster recovery capabilities are required? The answers to these questions will be unique to each organization.

- *Data Management*: Consider how the IoT solution handles data management, storage, processing, and analytics. Evaluate the scalability, performance, and flexibility of the data storage and analytics components to derive actionable insights, make informed decisions, and optimize processes based on IoT data.
- *Automation*: Another major strength of the IoT is the ability to automate devices using data, like the data collected by sensors. IoT automates various systems in your building by integrating smart devices like sensors and meters. This automation can reduce manual work and accelerate plumbing, electrical, and HVAC system functions. IoT offers the benefits of automation for different processes and operations in industries.
- *Green Buildings*: The IoT has become a favorite tool of designers of green buildings. IoT in engineering can help buildings reach higher standards of functionality, safety, and environmental responsibility. IoT can also create advanced structures that are safer and more environmentally friendly. In today's business landscape, investing in green technology is best for saving money and improving operations. When it comes to this, the IoT can be a viable solution.
- *Efficiency*: IoT systems, including smart thermostats and meters, offer precise energy control and monitoring. They can enhance efficiency and sustainability in any property. By regulating energy consumption, these systems not only reduce utility bills but also improve ambiance, occupant comfort, and overall operations. Investing in energy-efficient devices and systems is crucial. Cutting-edge IoT technologies like intelligent lighting and dynamic glass enhance your building's aesthetics and functionality, ultimately increasing its overall value. Smart buildings are not only highly functional but are visually attractive, as typically shown in Fig. 7.10 [1].

7.6 Challenges

Although IoT adoption is transforming industries like agriculture and manufacturing, the increased connectivity also brings significant security risks. For IoT in architecture to work properly, the mass connectivity of several devices must be carefully managed so they work seamlessly together. When implementing IoT applications in architecture, major challenges include security concerns due to the vast network of connected devices, lack of standardization across different protocols and devices, managing large volumes of data, ensuring reliable connectivity, power constraints for battery-operated sensors, and addressing privacy issues related to collected data. Other challenges include the following [15]:

- *Security*: Data collection is one of the features of IoT technology, however, managing this data can lead to data security and privacy challenges. Security is a critical consideration for IoT solutions to protect data, devices, and networks from cyber threats, unauthorized access, and privacy breaches. Evaluate the security

Fig. 7.10 Visually attractive smart buildings [1]

features and protocols implemented in the solution, including encryption, authentication, access control, secure bootstrapping, and firmware updates. Device-level security involves the integration of special safeguards in the firmware and hardware for IoT networks.

- *Privacy Concerns*: Privacy concerns refer to issues related to the collection, storage, use, and sharing of personal information. This can include concerns about who has access to personal information, how it is being used, and whether it is being protected from unauthorized access or misuse. In the digital age, privacy concerns have become increasingly important as personal information is being collected and stored on an unprecedented scale. To address privacy concerns, individuals and organizations need to implement appropriate security measures to protect personal information, be transparent about how it is being used, and respect individuals' rights to control their own information.
- *Zero Trust*: The need for zero trust IoT security is readily apparent when considering Industrial IoT (IIoT) use cases. Implementing zero trust IoT architecture is crucial for protecting networks by ensuring only secure, explicit connections between devices and resources. With only explicit connections from a user to a resource being allowed, IoT zero trust standards are critical for ensuring modern cellular and wide-area network (WAN) security. Zero trust architecture uses site-to-site encryption and can include IoT remote access functions to provide secure communication across sites, vehicles, devices, applications, users, and the cloud.

- *Lack of Standard*: There is lack of a single, unified standard for data communication and management. Without mandatory IoT standards, security continues to be an area for improvement for many organizations. Different manufacturers use diverse protocols and standards for their IoT devices, making integration difficult and complex.
- *Scalability*: Managing the increasing number of connected devices as an IoT system expands can be challenging. Consider how a potential IoT solution will accommodate future growth, including not only actual data volume and usage, but also the number of IoT devices and types of devices used, as well as the complexity of the applications used and analysis being conducted.
- *Interoperability*: It is essential to choose interoperable IoT solutions based on open standards, not closed proprietary technologies. Interoperability facilitates the integration of new sensors and connected devices as needs evolve. It provides more flexibility to evolve installations over time.
- *Power Management*: This involves optimizing battery life for battery-powered sensors, especially in applications requiring long-term monitoring. It also involves balancing data collection frequency with power consumption.
- *Vulnerability*: Vulnerability to network attacks refers to the susceptibility of a network, system or device to being compromised or exploited by cyber criminals. This can happen due to weaknesses in the network infrastructure, unpatched software, poor password management, or a lack of appropriate security measures. Network attacks can result in data theft, loss of privacy, disruption of services, and financial loss. To reduce vulnerability to network attacks, it's important to implement strong security measures such as firewalls, encryption, and regular software updates.

7.7 Conclusion

The Internet of things is a set of rapidly evolving technologies. It shows exciting promise for revolutionizing the functions of architecture, construction, and engineering (ACE) in several ways. It is changing the architecture and design trades. The IoT technology has revolutionized how we design and build modern buildings. The IoT in architecture, design, and construction will shape processes and the outcomes as we move into the future. More information about Internet of things in the architecture industry can be found in the books in [16, 17] and the following related journal: *IEEE Internet of Things Journal*.

References

1. IoT applications for smart buildings: Use cases and top benefits (2024). https://www.digi.com/blog/post/iot-applications-for-smart-buildings-use-cases-and

2. M.N.O. Sadiku, P.A. Adekunte, J.O. Sadiku, Internet of things in architecture. Int. J. Trend Sci. Res. Dev. **9**(1), 294–302 (2025)
3. M. Townes, The spread of TCP/IP: how the Internet became the internet. Millennium J. Int. Stud. **41**(1), 43–64 (2012)
4. Internet of things (IoT). https://www.geeksforgeeks.org/internet-things-iot-2/
5. M.N.O. Sadiku, S.M. Musa, S.R. Nelatury, Internet of things: an introduction. Int. J. Eng. Res. Adv. Technol. **2**(3), 39–43 (2016)
6. P. Sadeghi et al., Towards a reliable modulation and encoding scheme for Internet of things communications, in 13th IEEE International Conference on Application of Information and Communication Technologies (AICT) (2019). https://www.researchgate.net/figure/Communications-Technologies-in-Internet-of-Things_fig1_335104959
7. M.N.O. Sadiku, Y. Wang, S. Cui, S.M. Musa, Industrial internet of things. Int. J. Adv. Sci. Res. Eng. **3**(11), 1–4 (2017)
8. A.R. Sadeghi1, C. Wachsmann, M. Waidner, Security and privacy challenges in industrial Internet of things, in Proceedings of the 52nd Annual Design Automation Conference (2015)
9. The industrial Internet of things (IIoT): The business guide to Industrial IoT. https://www.i-scoop.eu/internet-of-things-guide/industrial-internet-things-iiot-saving-costs-innovation/
10. D. Schulz, FDI and the industrial Internet of things, in Proceedings of IEEE 20th Conference on Emerging Technologies & Factory Automation, pp. 1–8 (2015)
11. J. Rahm, Internet of things in the manufacturing industry (2017). https://blog.flexlink.com/internet-of-things-in-the-manufacturing-industry/
12. S. Shakeri, Top 7 IoT applications in architecture and design (2022). https://parametric-architecture.com/top-7-iot-applications-in-architecture-and-design/?srsltid=AfmBOoqjg-zqmBLzpm0gI_lqAPP8IBvG-B_oNZ_wguMHJjnCGyrgKpDG#google_vignette
13. How the Internet of things (IoT) is shaping architecture, design, and construction. https://largeformat.hp.com/au/blog/how-the-internet-of-things-iot-is-shaping-architecture-design-and-construction
14. IoT architecture explained. https://www.mongodb.com/resources/basics/cloud-explained/iot-architecture
15. Challenges in Internet of things (IoT) (2023). https://www.geeksforgeeks.org/challenges-in-internet-of-things-iot/
16. M.N.O. Sadiku, *Internet of Things and Its Applications* (Lambert Academic Publishing, Moldova, Europe, 2024)
17. P. Lea, *Internet of Things for Architects: Architecting IoT Solutions by Implementing Sensors, Communication Infrastructure* (Analytics, and Security. Packt Publishing, Edge Computing, 2018)

Chapter 8
Big Data in Architecture

Every company has big data in its future and every company will eventually be in the data business.—Thomas H. Davenport.

8.1 Introduction

Big data has been a buzzword in recent years. The increasing amount of data raises both the opportunities and the challenges of managing it. Figure 8.1 shows Google's data center in Mayes County, Oklahoma, USA, where data is stored [1]. When properly analyzed, big data can reveal patterns, trends, and relationships related to human behavior and interaction in the built environment. Today's organizations depend on data and insights to make most of their decisions.

Big data in architecture refers to the application of big data technologies and techniques within the field of architecture, where large volumes of data from building sensors, user interactions, and environmental factors are collected and analyzed to optimize building design, performance, and operations. It also refers to the extensive datasets collected from various technologies, such as IoT sensors, social media, and satellite imaging, which provide valuable insights into design practices and urban planning. In the architecture industry, big data can be utilized to analyze vast amounts of information like building usage patterns, environmental data, construction costs, and design trends, allowing architects to make data-driven decisions for more efficient, sustainable, and user-centric building designs.

Big data is a broad term for the collection of both structured and unstructured data. Data is one of the biggest byproducts of the twenty-first century. Almost everything we do produces data, from swiping credit cards to emailing. Our production of data is exploding. Increasingly, the value of a business is tied to its ability to mine data. In the field of architecture, data is having a similar impact. In the modern era of architecture, big data plays a crucial role in transforming design processes. Big data and architecture are a combination worth exploring in depth [2].

This chapter examines the role of big data in architecture. It begins with explaining what big data is all about. It describes some characteristics of big data. It explains big

M. N. O. Sadiku, *Emerging Technologies in Architecture*,
https://doi.org/10.1007/978-3-031-99509-5_8

Fig. 8.1 Google's data center [1]

data in architecture and provides some of its applications. It highlights the benefits and challenges of big data in architecture. The last section concludes with comments.

8.2 What is Big Data?

Big data applies to data sets of extreme size (e.g. exabytes, zettabytes) which are beyond the capability of the commonly used software tools. It involves situation where very large data sets are big in volume, velocity, veracity, and variability [3]. The data is too big, too fast, or does not fit the regular database architecture. It may require different strategies and tools for profiling, measurement, assessment, and processing.

Big Data is essentially classified into three types [4]:

- *Structured Data*: This is highly organized and is the easiest to work with. Any data that can be stored, accessed, and processed in the form of fixed format is known as a structured data. It may be stored in tabular format. Due to their nature, it is easy for programs to sort through and collect data. Structured data has quantitative data such as age, contact, address, billing, expenses, credit card numbers, etc. Data that is stored in a relational database management system is an example of structured data.

- *Unstructured Data*: This refers to unorganized data such as video files, log files, audio files, and image files. Any data with unknown form or the structure is classified as unstructured data. Almost everything generated by a computer is unstructured data. It takes a lot of time and effort required to make unstructured data readable. Examples of unstructured data include Metadata, Twitter tweets, and other social media posts.
- *Semi-structured Data*: This falls somewhere between structured data and unstructured data, i.e., both forms of data are present. Semi-structured data can be inherited such as location, time, email address, or device ID stamp.

The different types of big data are depicted in Fig. 8.2 [5].

The process of examining big data is often referred to big data analytics. It is an emerging field since massive computing capabilities have been made available by e-infrastructures [6]. Big data analytics is the application of advanced analytic techniques to large, heterogeneous data sets that comprise structured, semi-structured, and unstructured data from many sources with sizes ranging from terabytes to zettabytes.

Analytics include statistical models and other methods that are aimed at creating empirical predictions. Data-driven organizations use analytics to guide decisions at all levels. Several techniques have been proposed for analyzing big data. These include the HACE theorem, cloud computing, Hadoop, and MapReduce [7].

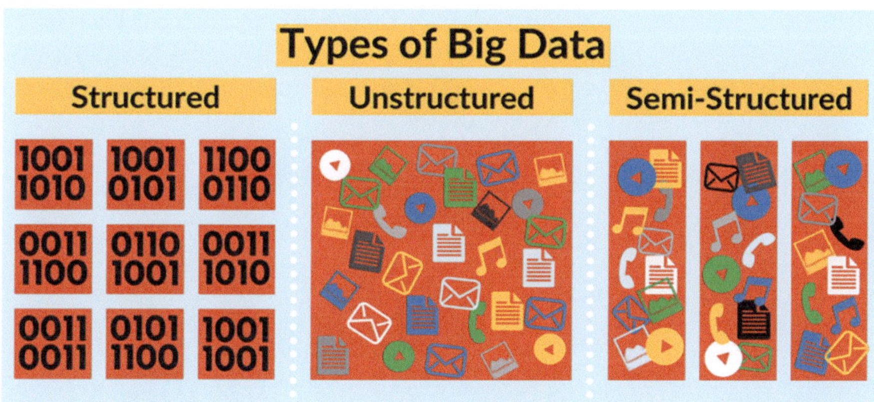

Fig. 8.2 Types of big data [5]

8.3 Characteristics of Big Data

Big data is growing rapidly and expanding in all science and engineering, including physical, biological, and medical services. Different companies use different means to maintain their big data. As shown in Fig. 8.3 [8], big data is characterized by 42 Vs. The first five Vs are volume, velocity, variety, veracity, and value.

- *Volume*: This refers to the size of the data being generated both inside and outside organizations and is increasing annually. Some regard big data as data over one petabyte in volume.
- *Velocity*: This depicts the unprecedented speed at which data are generated by Internet users, mobile users, social media, etc. Data are generated and processed in a fast way to extract useful, relevant information. Big data could be analyzed in real time, and it has movement and velocity.
- *Variety*: This refers to the data types since big data may originate from heterogeneous sources and is in different formats (e.g., videos, images, audio, text, logs). BD comprises of structured, semi-structured or unstructured data.
- *Veracity*: By this, we mean the truthfulness of data, i.e. whether the data comes from a reputable, trustworthy, authentic, and accountable source. It suggests the inconsistency in the quality of different sources of big data. The data may not be 100% correct.

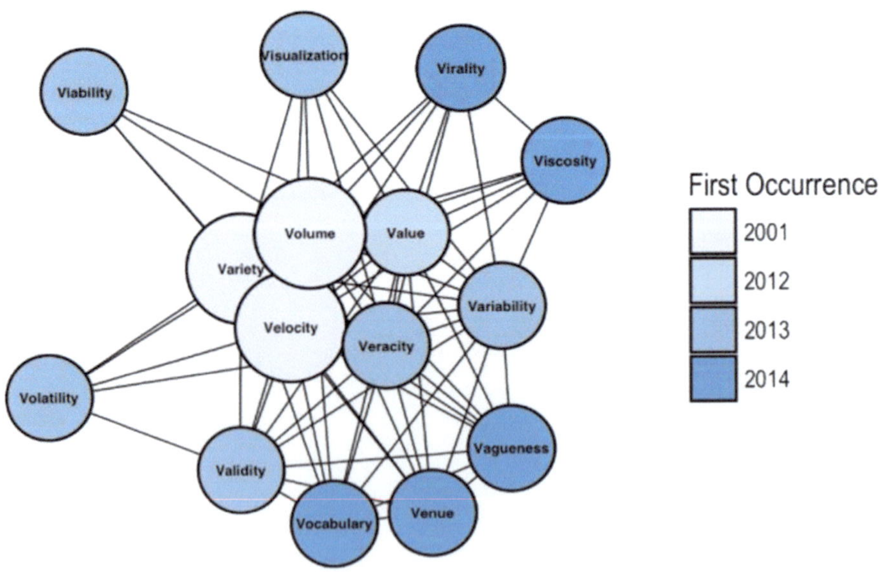

Fig. 8.3 The 42 V's of big data [8]

- *Value*: This is the most important aspect of the big data. It is the desired outcome of big data processing. It refers to the process of discovering hidden values from large datasets. It denotes the value derived from the analysis of the existing data. If one cannot extract some business value from the data, there is no use managing and storing it.

On this basis, small data can be regarded as having low volume, low velocity, low variety, low veracity, and low value. Additional five Vs have been added [9]:

- *Validity*: This refers to the accuracy and correctness of data. It also indicates how up to date it is.
- *Viability*: This identifies the relevancy of data for each use case. Relevancy of data is required to maintain the desired and accurate outcome through analytical and predictive measures.
- *Volatility*: Since data are generated and change at a rapid rate, volatility determines how quickly data change.
- *Vulnerability*: The vulnerability of data is essential because privacy and security are of utmost importance for personal data.
- *Visualization*: Data needs to be presented unambiguously and attractively to the user. Proper visualization of large and complex clinical reports helps in finding valuable insights.

Instead of the 10V's above, some suggest the following 5V's: Venue, Variability, Vocabulary, Vagueness, and Validity) [10].

Industries that benefit from big data include the healthcare, financial, airline, travel, restaurants, automobile, sports, agriculture, and hospitality industries. Big data technologies are playing an essential role in farming: machines are equipped with sensors that measure data in their environment. Structured and unstructured data are generated in various types [11–13].

8.4 Big Data in Architecture

Big data in architecture and design leverages vast datasets to enhance and personalize user experiences, optimize product functionalities, and streamline design processes through data-driven insights. An increasing number of gadgets in the built environment, such as thermostats and refrigerators, are bolstering the Internet of things and relaying the data that they gather. Data has even manifested a physical presence. In New York, a new type of architecture is emerging in which large skyscrapers are being retrofitted into digital warehouses that accommodate computers rather than people. An example is 375 Pearl Street in New York, shown in Fig. 8.4 [1]. Similar buildings are popping up across the United States for the purpose of storing and analyzing data.

When architects embark on a project, they gather data from a variety of sources, most often directly from the client. They determine which data points will help them

Fig. 8.4 A building in New York for storing and analyzing data [1]

create buildings that will have the greatest impact on their end users. They also tailor data analysis to every new project. Architects must gather as much data as possible in the early stages of the design process, because even small details can impact the efficiency of an entire building.

Data sources for big data in architecture include data on HVAC systems, lighting, and other building systems such as connected devices like smart meters and wearable technology. It also includes environmental sensors that monitor air quality, temperature, humidity, and other environmental parameters. Key aspects of big data in architecture:

Data collection: Sensors embedded in buildings gather data on energy consumption, temperature, humidity, occupancy levels, lighting conditions, and more. Data collection is now part of our daily lives, from the temperature set on our thermostat to the time we walk down the sidewalk.

Data analysis: Advanced analytics tools are used to process and interpret the collected data, identifying patterns and trends to inform design decisions.

Data integration: Combining data from various sources (sensors, building management systems, weather data) can be complex.

Data is changing architecture in the following three ways [1]:

1. *Clients are demanding data from architects*: Clients are starting to ask architects to deliver more than just drawing sets. They are eyeing the data-rich BIM models that companies use to document projects as a way to supply data for downstream applications, such as facilities management. For architects, this means that their data needs to be as rigorous as their drawings.
2. *Clients are demanding data from buildings*: Clients have also become interested in the data generated by the buildings. This data enables building owners to measure and improve their facilities' performance quantitatively. Architects also need to recognize that clients are going to use this data to measure their own performance. Performance-based contracts, where a portion of an architect's fee is withheld until post-occupancy data validates its prescribed design performance, are gaining popularity.
3. *Data is changing the process as much as it is changes the output*: The abundance of data may give rise to data warehouses, but the much more profound changes for architects will be procedural. If architects are to harness data from the built environment, even more significant procedural changes may be coming.

8.5 Applications of Big Data in Architecture

In architecture, big data applications can be used for various purposes including facility performance monitoring, predictive maintenance, energy optimization, space utilization analysis, design optimization through data-driven insights, construction project management, and even urban planning. Common applications of big data in architecture include the following [14]:

- *Building Performance Monitoring* (BPM): Sensors deployed throughout a building collect data on energy consumption, temperature, humidity, CO2 levels, and occupancy patterns, which can be analyzed to identify areas for improvement and optimize energy usage. This involves analyzing energy usage data to identify areas for improvement, optimizing HVAC systems, and adjusting lighting based on occupancy. It may monitor energy usage, indoor air quality, and occupant comfort levels in real-time to identify areas for improvement. Real-time data can be used to optimize building performance while studying occupants can be useful for user-centered design.
- *Predictive Maintenance*: By analyzing sensor data from building systems, potential failures can be predicted before they occur, allowing for proactive maintenance and preventing costly breakdowns. This may entail identifying potential issues in building systems before they occur, preventing costly breakdowns.

- *Space Utilization Analysis*: Data from occupancy sensors and building management systems can be used to understand how spaces are being used, leading to improved space planning and allocation.
- *Design Optimization*: Analyzing large datasets on climate patterns, building materials, and user behavior can inform design decisions to create more sustainable and user-friendly buildings. This may also involve analyzing user behavior to optimize building layout, lighting, and ventilation for occupant comfort. It may identify potential delays and optimize construction schedules using data from past projects.
- *City Planning*: Cities use big data to improve urban planning and urban development. Analyzing demographic data, traffic patterns, and environmental factors can help urban planners design more sustainable and resilient cities. This data-driven approach to urban planning can guide decision-making processes not only for architects and urban planners but also for city officials to create policies for more livable cities. New York City, for example, developed an initiative known as NYC Open Data, a public platform that provides over 1,300 datasets from housing and construction, education, health, and safety, to transportation. Figure 8.5 shows Manhattan in New York [14]. Another example is MX3D Bridge, a smart 3D-printed bridge in the City of Amsterdam to analyze pedestrian and crowd behavior. The smart bridge uses a sensor network to collect data from its users. The goal is to predict future behaviors to improve its 3D-printed manufacturing technique.

Fig. 8.5 Manhattan in New York [15]

Fig. 8.6 A synthetic wall design [15]

- *Smart City Design*: Heavily connected "smart cities" are becoming more prominent in society and are no longer the stuff of future-oriented dreams. Although some of them are built from scratch, other city officials bring in architects and similar professionals. Although smart cities rely on architects to design physical buildings, they also depend on data architects to develop big data-based infrastructures that coincide with the easily visible structures. Smart cities use big data for advertising purposes. They use data from traffic, pollution, and transport to optimize urban planning.
- *Synthesis Design*: Big data analytics was used to design an interior feature wall for the Watson Experience Center in San Francisco. They used data from the influence of mobile phones on monthly consumer spending to create a precise screen material that defines the wall. Figure 8.6 shows a synthetic wall design [15]. Without any explanation, the design gives little indication that it represents a massive data set.

8.6 Benefits

The application of big data in architecture brings numerous benefits that significantly influence the quality and efficiency of projects. In architecture, big data is used to improve design decisions, operational efficiency, and overall building performance.

Integrating big data into architecture can lead to more informed, sustainable, and efficient design solutions. By collecting and analyzing data on past projects, construction companies can use predictive analytics to identify potential risks. Other benefits of big data in architecture include the following [14]:

- *Clash Detection*: With the help of big data, we can anticipate potential design problems before construction begins. For example, using building information modeling (BIM) it is now common practice to do collision detection to check all areas. This reduces costs for our clients and much less headaches for architects as there are fewer design changes during construction.
- *Energy Efficiency*: Another area where big data shines is improving energy efficiency. Using data sets of weather patterns, solar trajectories, wind direction, and strength, we can design better buildings that maximize natural light, ventilation, and heating for a more sustainable future.
- *Project Management*: Tracking progress on construction sites using sensors and data analytics can help identify potential delays, optimize resource allocation, and improve overall project efficiency.
- *Parametric Design*: Parametric design in architecture involves the use of parameter-based algorithms to design buildings that automatically adapt to changing variables, allowing for customizable and flexible structures.
- *Enhanced Decision-Making*: Data provides the foundation for evidence-based design decisions. Big data is transforming architectural design processes by enabling more informed decision-making through data-driven insights. It allows architects to analyze user behavior, environmental conditions, and urban trends, improving design accuracy and sustainability. The data-driven approach facilitates informed decision-making, reducing waste, and enhancing sustainability in architecture.
- *Anticipate the Future*: The architects of the past had to largely rely on guesswork when calculating things like population growth or the effects of climate change. Big data aids architects in understanding places and people, then incorporating the associated revelations into their designs. Datasets and modeling could help architects test the viability of design concepts before constructing them in real life.
- *Waste*: Waste is a major practical and political concern. It also becoming an increasingly public issue as news outlets publish pieces about pollution, plastics in the ocean, and whales killed by car parts. Figure 8.7 shows waste [16]. Recycling is viewed as the responsible way of dealing with waste. By embracing big data, however, businesses can streamline their disposal practices where it really matters. Data systems all play a role in moving the US towards being a zero waste system.
- *Sustainable Design*: Utilizing data to design buildings that minimize environmental impact through energy efficiency and resource optimization. Big data informs sustainable architectural design by analyzing environmental impacts, optimizing resource use, and predicting energy efficiency. It allows architects

Fig. 8.7 Waste [16]

to assess site conditions, climate, and material performance to create eco-friendly buildings.

8.7 Challenges

Although big data has tremendous impact on industries, it is not without challenges. The challenges of integrating big data into architectural design include managing and analyzing large volumes of data, ensuring data privacy and security, the need for specialized skills and technology, and bridging the gap between data insights and creative design processes. Opting for a big-data enabled data analytics solution is not straightforward. It requires vast technology land space for components to ingest data from numerous sources. It is also essential to have proper synchronization between these components. Keeping up with varying use cases of big data is a significant challenge for many organizations. Other challenges in using big data in architecture include the following [16]:

- *Data Security*: Although big data can provide great insight for decision-making, protecting data from theft is challenging.
- *Data Privacy*: Data collected may contain personal information. It is necessary to ensure user data is protected when collecting information about building occupants.

- *Data Visualization*: Presenting complex data insights in a clear and understandable way for architects and building owners.
- *Data Integration*: Combining data from various sources can be complex.
- *Data Quality*: Ensuring the accuracy and reliability of data is crucial.
- *Data Interpretation*: Another challenge is the interpretation and application of data. The architect are liable for making the final decision. Misinterpretation can lead to wrong decisions, leading to ineffective architectural solutions.
- *Data Storage*: Data volume remains a significant challenge because data volumes are doubling in size about every two years. Besides data size, the number of file formats used to store data is also growing. As a result, effectively storing and managing information is often a challenge for an organization.
- *Data Analytics*: Data analysis in architecture is a complicated process. Design teams must not only gather the most accurate data, but also analyze it and locate hidden patterns that predict how people will most efficiently use a building.
- *Data Quality*: Data quality aspects include accuracy, consistency, relevance, completeness, and use fitness. Data quality is a challenge anytime working with diverse data sources. Consequently, obtaining useful data requires a significant effort to clean the data before bringing it for analysis.
- *Learning New Data Tools*: Despite its transformative potential, it can be challenging for firms to incorporate big data into architecture due to the sheer volume and complexity of data. Architects often do not have the skill set required to understand complex data and algorithms. This means they have to constantly learn new tools to implement big data in our practice.
- *Scaling*: Big data solutions are used to handle large volumes of data. It can cause issues if the planned architecture is unable to scale. This may degrade the application performance and efficiency.
- *Complexity*: Big data systems can be challenging to implement since they must deal with various data types from various sources.
- *Collaboration*: This is essential during the design process, especially when architects and their clients gather data simultaneously. It is crucial to use data analysis tools that foster collaboration between all parties involved in the building's development.
- *Skillset*: Big data technologies are highly specialized, and they use frameworks and languages that are not common in more general application architectures. To operate these modern technologies and data tools, skilled data professionals are required. These will include data scientists, analysts, and engineers to operate tools and get data patterns. A shortage of data experts is one of the challenges that companies face.
- *Emerging Technologies*: Emerging technologies such as artificial intelligence, machine learning, and Internet of things are expected to play an increasingly important role in analyzing and interpreting big data. Architects need to learn about these technologies and harness the power of big data to become better designers of buildings and cities.

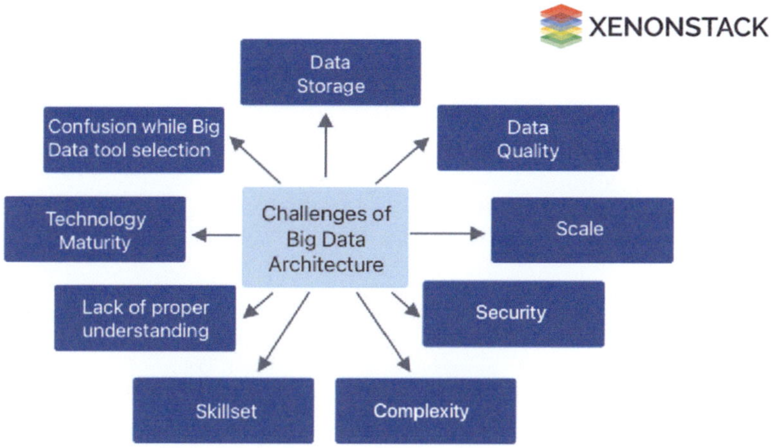

Fig. 8.8 Some of the challenges of big data in architecture [17]

- *Cost*: This can also be a challenge when performing data analysis in architecture. Third-party data formatting can be a long, expensive process, depending on the scope of the project.

  Some of these challenges are displayed in Fig. 8.8 [17].

8.8 Conclusion

Data analytics in architecture is vital. Data analytics in architecture can be used to improve wayfinding at a facility. But in order to use data effectively, we must first understand what types of data to gather and how to analyze it properly. While integrating big data into architecture promises impressive benefits, it also brings unique challenges. Despite the challenges, big data is here to stay. Those who can harness big data will undoubtedly lead the charge toward more responsive and responsible built environments. More information about big data in the architecture industry can be found in the books in [18, 19].

References

1. D. Davis, How big data is transforming architecture (2015). https://www.architectmagazine.com/technology/how-big-data-is-transforming-architecture_o
2. M.N.O. Sadiku, U.C. Chukwu, J.O. Sadiku, Big data in architecture. Innov. Mult-Discip. J. Appl. Technol. **3**(1), 28–39 (2025)
3. M.N.O. Sadiku, M. Tembely, S.M. Musa, Big data: an introduction for engineers. J. Sci. Eng. Res. **3**(2), 106–108 (2016)

4. The complete overview of big data. https://intellipaat.com/blog/tutorial/hadoop-tutorial/big-data-overview/
5. R. Allen, Types of big data|Understanding & Interacting with key types (2024). https://investguiding-com.custommapposter.com/article/types-of-big-data-understanding-amp-interacting-with-key-types
6. P. Baumann et al., Big data analytics for earth sciences: the earthserver approach. Int. J. Digit. Earth **19**(1), 3–29 (2016)
7. X. Wu et al., Knowledge engineering with big data. IEEE Intell. Syst. 46–55 (2015)
8. The 42 V's of big data and data science. https://www.kdnuggets.com/2017/04/42-vs-big-data-data-science.html
9. P.K.D. Pramanik, S. Pal, M. Mukhopadhyay, in *Healthcare Big Data: A Comprehensive Overview*, ed. by N. Bouchemal (ed.), Intelligent Systems for Healthcare Management and Delivery (IGI Global, chapter 4, 2019), pp. 72–100
10. J. Moorthy et al., Big data: prospects and challenges. J Decis Makers **40**(1), 74–96 (2015). https://www.grandviewresearch.com/industry-analysis/industrial-wireless-sensor-networks-iwsn-market
11. A.K. Tiwari, H. Chaudhary, S. Yadav, A review on big data and its security. in Proceedings of IEEE Sponsored 2nd International Conference on Innovations in Information Embedded and Communication Systems (2015)
12. M.B. Hoy, Big data: an introduction for librarians. Med. Ref. Serv. Q. **33**(3), 320–326 (2014)
13. M. Viceconti, P. Hunter, R. Hose, Big data, big knowledge: big data for personalized healthcare. IEEE J. Med. Health Inf. **19**(4), 1209–1215 (2015)
14. The promises and pitfalls of big data in architecture (2024). https://slantis.com/blog/the-promises-and-pitfalls-of-big-data-in-architecture
15. Big data becomes architecture in this CNC-milled screen wall for IBM (2017). https://www.archdaily.com/870095/big-data-becomes-architecture-in-this-cnc-milled-screen-wall-for-ibm
16. L. Alton, Facing environmental crisis, data is reshaping waste management. https://www.smartdatacollective.com/facing-environmental-crisis-data-is-reshaping-waste-management/#google_vignette
17. C. Gaur, Top 9 challenges of big data architecture and its solutions|Overview (2023). https://www.xenonstack.com/insights/challenges-of-big-data-architecture
18. M.N.O. Sadiku, U.C. Chukwu, P.O. Adebo, *Big Data and Its Applications* (Lambert Academic Publishing, Moldova, Europe, 2024)
19. S.M.F. Akhtar, *Big Data Architect's Handbook: A Guide to Building Proficiency in Tools and Systems Used by Leading Big Data Experts* (Packt Publishing, 2018)

Chapter 9
Cloud Computing in Architecture

> *Cloud computing offers individuals access to data and applications from nearly any point of access to the Internet, offers businesses a whole new way to cut costs for technical infrastructure, and offers big computer companies a potentially giant market for hardware and services.—Jamais Cascio.*

9.1 Introduction

Architecture is the art/science of creating solutions that map to business needs. The architecture, engineering, and construction (AEC) sector is a highly fragmented, data intensive, project based industry, involving a number of very different professions and organizations. Systems and standards are continually evolving in the realm of AEC and often serve as the catalyst for innovation. Cloud computing is the newest technology that delivers computing services over the web. It allows users to access and use resources on demand, without having to invest in or maintain physical infrastructure. It also allows users to access additional storage, enhanced processing power, and hardware-heavy architecture software all remotely through the Internet. Moreover, the architecture industry is under increasing pressure to deliver high-quality projects in shorter timeframes, leaving no room for human error. Cloud platforms allow architects to fix mistakes quickly. Architects utilize cloud computing primarily to streamline their design process by enabling faster rendering of 3D models, real-time collaboration on projects, and storing large design files on remote servers, all while reducing the need for expensive local hardware and software licenses.

Architects are increasingly using cloud computing to access and manage design software, collaborate on projects in real-time with team members across different locations, store large design files, and perform computationally intensive tasks like 3D rendering, all through a centralized online platform, significantly enhancing efficiency and collaboration within their workflow. Architects today are starting to use the cloud for more than just file sharing and storage [1]. Figure 9.1 shows an architect at work [2].

M. N. O. Sadiku, *Emerging Technologies in Architecture*,
https://doi.org/10.1007/978-3-031-99509-5_9

Fig. 9.1 An architects at work [2]

Cloud computing (CC) is the delivery of computing services, such as servers, storage, databases, networking, software, analytics, and intelligence, over the Internet. It is no longer a business trend. It is a new reality in the global market. The power of cloud computing lies in the fact that it is available anywhere, in any size or configuration, and can be easily reconfigured when needed. Using this kind of decentralized system offers flexibility, scalability, and accessibility for architectural teams all around the globe. Cloud computing is emerging as the cornerstone of a new era in architecture; it transcends conventional boundaries and unleashes unparalleled possibilities in the field. It can synchronize architectural projects from concept to construction. Migrating to the cloud enables architects to store, manage, and process their data online [3].

This chapter explores ways cloud computing helps organizations in the architecture industry become more efficient, sustainable, and innovative. It begins with explaining the basics of cloud computing. It describes cloud computing in architecture and provides some of its applications. It highlights the benefits and challenges of cloud computing in architecture. The last section concludes with comments.

9.2 Cloud Computing Basics

Cloud computing represents a newly emerging service-oriented computing technology. It is the provision of scalable computing resources as a service over the Internet. It allows manufacturers to use many forms of new p systems such as 3D printing, high performance computing (HPC), industrial Internet of things (110T), and industrial robots. It is transforming virtually every facet of modern manufacturing. It is innovating, reducing cost, and bolstering the competitiveness of American manufacturing [4].

The key characteristic of cloud computing is the virtualization of computing resources and services. Cloud computing is implemented in one of three major formats: software as a service (SAAS), platform as a service (PAAS), or infrastructure as a service (IAAS). These services are illustrated in Fig. 9.2 [5] and explained as follows:

SaaS: This is a software delivery model in which software and associated data are hosted on the cloud. In this model, cloud service providers offer on-demand access to computing resources such as virtual machines and cloud storage. Nowadays oil & gas companies transition to cloud computing and implement SaaS solutions for operations.

PaaS allows the end-user to create a software solution using tools or libraries from the platform service provider. In this model, cloud service providers deliver computing platforms such as programming and execution.

In the laaS model, cloud service providers can rent manufacturing equipment such as 3D printers.

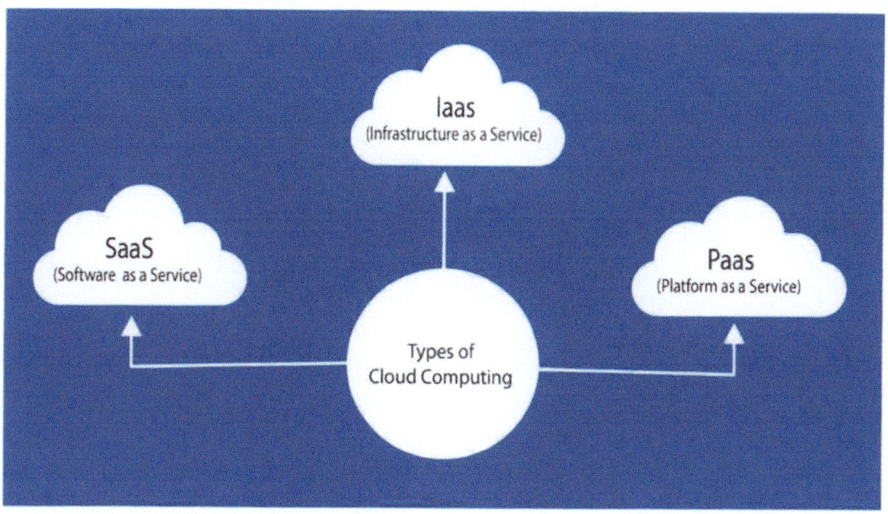

Fig. 9.2 Three types of cloud computing services [5]

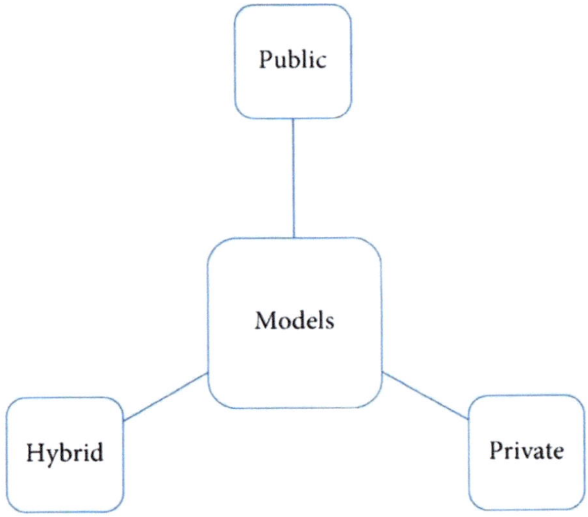

Fig. 9.3 Cloud computing models [7]

Just like cloud computing, CM services can be categorized into three major deployment models (public, private, and hybrid clouds) [6]:

- Private cloud refers to a centralized management effort in which manufacturing services are shared within one company' or its subsidiaries. A private cloud is often used exclusively by one organization, possibly with multiple business units.
- Public cloud realizes the key concept of sharing services with the general public. Public clouds are commonly implemented through data centers operated by providers such as Amazon, Google, IBM, and Microsoft.
- Hybrid cloud that spans multiple configurations and is composed of two or more clouds (private, community or public), offering the benefits of multiple deployment modes.

These models are shown in Fig. 9.3 [7]. Cloud computing finds application in almost every field.

9.3 Cloud Computing in Architecture

Although cloud computing (CC) is nearly ubiquitous on a personal level, for architecture companies, its adoption has been much slower. This can be attributed to the fact that an architectural project usually involves synchronizing the efforts of many professionals and the level of organization needed is much more sophisticated. The last few years have seen a massive shift in architects adopting digital capabilities from local servers to cloud computing. Cloud computing is turning out to be the connective tissue for more and more modern architectural teams.

Any firm of architects requires several fundamental tools and technologies to support productivity. Traditionally, computing for architects was almost entirely on-premises. All hardware was in the office, and software ran locally on in-office workstations. It was not too long ago when architects made the shift from manual drafting tables to digital canvases in the form of CAD. Today, one of the newest technological developments is cloud computing. With the emergence of cloud computing in the last 15 years, the way many industries work has been transformed [8].

In architecture, cloud computing refers to the design and implementation of a system where computing resources like servers, storage, and networking are delivered as a service over the Internet. This allows users to access and utilize these resources on-demand without managing the underlying infrastructure.

Shifting from traditional to cloud-based workflows requires a strategic approach. Choosing a cloud provider is of paramount importance. Architects need to consider factors like the service's reliability, security protocols, and integration capabilities. Picking the right service that aligns with specific architectural requirements establishes the groundwork for a smooth and successful transition to the cloud. For a small firm, the clear advantage is that a cloud computing subscription is much cheaper than running their own servers.

Cloud computing can significantly advance your career in architecture by enabling you to design more innovative and efficient buildings, collaborate seamlessly with remote teams, access powerful computational tools for complex analysis, manage large datasets efficiently, and showcase your designs through immersive virtual reality experiences. You do all this while staying at the forefront of technological advancements in the field. The best architects in cloud computing tend to be adept at most of what the platform has to offer. Figure 9.4 shows some cloud architects [2]. Much like an architect who designs buildings, a cloud architect designs virtual environments and infrastructures. A cloud architect is essentially an IT professional who designs, builds, and manages a company's cloud infrastructure.

9.4 Applications of Cloud Compting in Architecture

Cloud computing is providing compute power and other related software services on rent. In architecture, cloud computing is primarily used for collaboration, data sharing, data storage and analysis, project management tools, BIM (building information modeling) integration, visualization tools, and disaster recovery. CC enables architects to access and share large design files, collaborate remotely, and leverage data analytics for better decision-making throughout the design and construction process. Common examples applications of cloud computing in architecture include the following [9, 10]:

- *BIM Integration*: Cloud platforms can integrate with BIM software, allowing architects to upload and share BIM models for analysis, coordination, and clash detection with project stakeholders. This involves storing and managing large

Fig. 9.4 Some cloud architects [2]

BIM models on a cloud platform for collaborative design and analysis across teams. Cloud-based BIM software allows architects to store and access large project models from anywhere, facilitating better project coordination and data sharing. Storing and managing large BIM models on cloud platforms allows multiple team members to access and collaborate on design details simultaneously, facilitating real-time updates and coordination. Cloud storage services like Amazon S3 or Microsoft Azure Blob Storage enable efficient storage and access to large BIM models, facilitating easier collaboration and project version control. Figure 9.5 shows various components of BIM [11].

- *Design Visualization*: The majority of an architect's work revolves around producing documents, while renderings and drawings are the bread and butter of their work. Cloud-based rendering services provide high-quality visualization capabilities, enabling architects to create realistic 3D renderings and virtual tours quickly. Cloud-based platforms with advanced analytics capabilities can be used to analyze large datasets related to building performance, energy consumption, and environmental impact, providing valuable insights for design optimization. Utilizing cloud-based rendering services to produce high-quality visualizations and animations quickly, even for complex models, without requiring expensive local rendering hardware. Cloud-based rendering services allow architects to generate high-quality renderings and visualizations quickly, without requiring expensive hardware, facilitating better client presentations. Cloud platforms offer

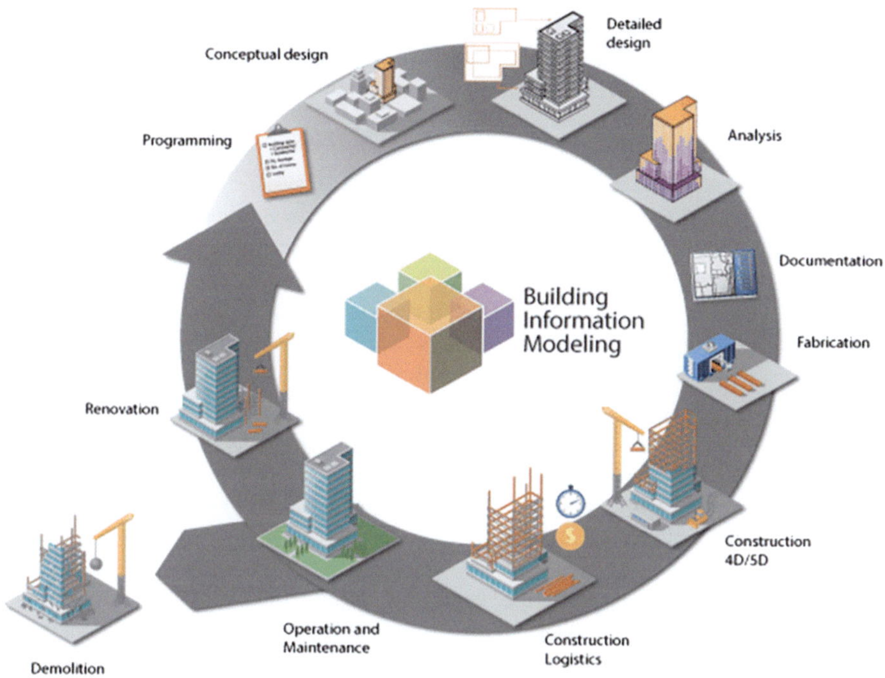

Fig. 9.5 Components of BIM [11]

high-performance computing power for rendering complex 3D models significantly faster, allowing architects to create high-quality visuals with minimal wait time.

- *Collaboration*: Cloud-based design software enables multiple team members to work on the same project simultaneously, making real-time collaboration and design feedback easier, regardless of location, facilitating real-time collaboration on design models and drawings. It allows storing project documents like drawings, specifications, and reports on a cloud platform for easy access and real-time updates by all team members. Its decentralized nature eliminates the need for teams to be in the same physical office to function. Having access to this type of seamless teamwork and remote access enables real-time collaboration and allows teams scattered around the globe to contribute to a project as if they were working in the same room.
- *Remote Access*: Before remote work became common, architects usually kept their project data on computers in their offices. While this could manage 3D models well, documents and spreadsheets can be better handled by cloud storage. Cloud computing allows teams to access and work on project information remotely. Architects and designers can access project files from anywhere with an internet connection, enabling remote work and collaboration. In architectural design, cloud computing allows architects to utilize remote servers to access powerful

processing capabilities for tasks like 3D modeling, rendering, design collaboration, and data storage. This results in faster design iterations, improved efficiency, and easier client communication, all while minimizing the need for expensive local hardware. Working remotely means that less time waste, resource waste, and carbon emissions are unnecessarily consumed by commuting to and from the office.

- *Data Analysis*: Cloud computing can be used to run complex simulations like energy modeling, structural analysis, and daylight studies on large datasets to optimize designs. Cloud-based analytics tools can process large datasets related to site conditions, energy usage, and building performance, helping architects make data-driven design decisions. They allow accessing powerful cloud computing resources to run complex energy simulations, structural analysis, and other design calculations.
- *Project Management*: This involves utilizing cloud-based project management platforms to track tasks, deadlines, budgets, and team communication across the project lifecycle. Cloud-based project management platforms like Asana or Trello can be used to streamline project workflows, track tasks, manage timelines, and communicate with team members effectively.
- *Disaster Recovery*: Cloud storage provides a reliable backup solution for critical design data, ensuring that projects are protected against data loss in case of hardware failures or natural disasters. Cloud storage and backup services provide a reliable way to safeguard critical design data against hardware failure or other disasters.
- *Construction Site*: Perhaps the biggest challenge of a building project is that the people who designed it are almost always located away from where it is being built. Even if a project is in their city, travel back and forth to the site is a huge time commitment for an architect. Unscheduled visits are often made to resolve single issues that require nothing more than designers seeing what the builder is talking about. An upside to using cloud computing in building projects is that the level of integration of CC allows a project team to develop their own custom tools. A typical construction site is shown in Fig. 9.6 [10].

9.5 Benefits

There are plenty of advantages gained by architects using CC. Cloud computing offers many advantages for architects, such as enhanced collaboration, increased efficiency, improved creativity, and greater innovation. Cloud computing tools are convenient, flexible, scalable, and cost-efficient. They suit any business type and size, from tiny agencies with a few workers to huge international enterprises. The benefits of enhanced collaboration, scalability, and digital security paint a compelling picture of the future of architectural design. Cloud computing is now sophisticated enough to synchronize architectural projects from concept to construction. Other benefits of using cloud computing in architecture include the following [9]:

Fig. 9.6 A typical construction site [10]

- *Improved Collaboration*: CC enables real-time collaboration on projects across geographically dispersed teams. Teams from different locations can work on the same project simultaneously. Collaborative design teams, dispersed globally, seamlessly can work on a project using cloud-based tools. Cloud-based design software like BIM (building information modeling) platforms allow multiple architects and stakeholders to work on the same project simultaneously, making real-time edits and accessing the latest design updates from anywhere with an Internet connection.
- *Cost Reduction*: Using cloud-based applications reduces rent and premises costs. CC eliminates the need for expensive high-end hardware and software licenses by leveraging cloud-based resources on a pay-per-use model. By utilizing cloud services, architects can avoid high upfront costs associated with powerful hardware and software licenses. One pays only for the resources used. Cloud computing allows architects to scale their computing resources as needed, reducing the need to invest in expensive hardware upfront.
- *Accessibility*: CC provides access to design data from any device with an Internet connection. Architects can access project data from anywhere with an internet connection. Redundancy allows ensuring continuous access to data and services.
- *Efficiency*: Using the cloud may significantly boost a company's efficiency. Faster rendering times, real-time collaboration, and easy access to design tools can significantly improve design workflow. Faster rendering, collaborative design, and streamlined workflows lead to improved project timelines and productivity.
- *Scalability*: CC allows for easy scaling of storage and computing power as project needs change. Accessing computing power based on project demands allows for efficient resource allocation. It is easy to scale computing resources up or down

based on demand. Architects can easily scale up computing power based on project needs, eliminating the need to invest in expensive hardware upgrades.

- *Flexibility*: There is access resources from anywhere with an Internet connection. Cloud computing provides architects the flexibility to adapt to project sizes and building typologies dynamically. Evolving projects can have the resources allotted to them be easily scaled up or down.
- *Data Security*: Cloud providers offer robust security features to protect sensitive design data. In the digital realm, ensuring security is essential. This is doubly important for architecture offices and the files they handle for their different projects. Cloud computing solutions excel in this regard, in how they implement advanced encryption, authentication protocols, and stringent access controls.
- *Sustainability*: The eco-friendly facets of digital architecture enabled by cloud solutions lie in how it decentralized the architectural design office. Exploring how cloud computing technologies contribute to sustainable practices will inevitably help in aligning the architectural industry with a commitment to environmental responsibility.
- *Faster Rendering*: Cloud platforms provide high-performance computing power to render complex 3D models significantly quicker than on local machines, saving time and allowing for more design iterations.
- *Visualization*: Clients can easily view and interact with 3D models through web browsers without needing specialized software installed on their computers, facilitating better communication and feedback loops.
- *Data Storage*: Cloud storage services provide a secure and scalable way to store large design files, project documents, and backups, eliminating concerns about local storage limitations. Cloud storage services like Amazon S3 or Azure Blob Storage enable architects to store and access massive amounts of design data (including 3D models, point clouds, and images) from anywhere, eliminating the need for large local storage systems. With cloud platforms, architects can organize large amounts of data with flexible storage options that do not affect budgets.
- *Complex Simulations*: Access to high-performance computing in the cloud enables architects to run computationally intensive simulations like energy modeling, structural analysis, and daylighting studies.
- *Design Competitions*: Cloud-based platforms can be used to host online design competitions, allowing architects to easily share their proposals with clients and reviewers globally.

9.6 Challenges

There are a couple of common challenges for cloud-based and remote computing among architects. Architects work on very large files. If architects need to edit them remotely, any lag or delay can be very frustrating. An architect needs to develop certain skills to take advantage of cloud computing, such as cloud fundamentals, cloud platforms, cloud applications, and cloud development. Continuous learning and staying updated with industry trends are essential in this dynamic field. Other challenges include the following [12]:

- *Security*: Working in the cloud enhances data security because it reduces the chances of accidental data loss or hacking of on-premise devices. A hacker is shown in Fig. 9.7 [13]. Data breaches can have significant financial and reputational costs. So architects should consider moving to cloud platforms whenever possible. Technologies such as Google Drive and DropBox are often used informally and in an ad-hoc way between individuals, but concerns over security and the protection of intellectual property often dissuade major companies from adopting such services.
- *Culture*: There is a significant cultural barrier to using the cloud in architecture, engineering, and construction. A lot of the consultants and architects are still working using AutoCAD. People are reluctant to change.
- *Transitioning to the Cloud*: Shifting from traditional to cloud-based workflows requires a strategic approach. People resist change. A step-by-step transition plan ensures a smooth integration, minimizing disruptions and maximizing the benefits of cloud computing. Architects can gradually migrate projects, ensuring a seamless evolution in their design processes.
- *Training*: The biggest skill you can have in the cloud space is willingness to learn. Providing architectural teams with essential skills is crucial for success. This is especially true with transitioning to a relatively new technology such as cloud computing which requires a new set of skills and know-how to navigate through. Learning or training on cloud computing can be done in a variety of ways. For example, you can take online courses, such as Coursera.

Fig. 9.7 A hacker [13]

- *Emerging Technologies*: The future holds plenty of potential in emerging technologies in architecture that can further be integrated into architectural cloud computing. For example, there is continually evolving realm of artificial intelligence for architecture, which automates parametric design programs. The answer to being prepared for any challenges in the field lies in mastering these new technologies.
- *Democratizing Design Work*: This involves using design platforms in the cloud where everyone can see drawings and collaborate on them. This "democratizes" architectural practice. It means firms no longer must rely on a hierarchical approach where one person knows everything, and other employees cannot participate in the creative process.
- *Interoperability*: Projects carried out within AEC sector involve collaboration between various people, using a variety of different systems. This, along with the industry's strong data sharing and processing requirements, means that the management of building data is complex and challenging.
- *Data Management*: Data management in the architecture industry can often be fragmented with a lack of an overall data management policy. Data sets relating to a particular project can often be stored in local computers of designers/architects, often with limited network connectivity, persistence, and availability.
- *Data Processing*: This is also an important concern for the architecture industry. During construction, a large proportion of work takes place on construction sites where computing resources are limited. Allowing users to make changes on a portable device on site that can then be processed remotely leading to the plans on site being updated is extremely desirable.
- *Governance of BIM Data*: BIM allows a building to be modelled across its entire life-cycle, from concept design to construction. Cloud computing capability would make most sense when utilized alongside a BIM data representation. However, due to the complex project based nature of the AEC industry, any data stored in a cloud system would need to be heavily managed. This level of management is essential to ensure that the data is able to meet the legal and contractual requirements for each individual project and to also ensure that the intellectual property rights.
- *Storage*: One of the key issues within the industry is the storage of building data during design/construction and over the entire life of the building. Several companies have developed servers for the storage of building data.

9.7 Conclusion

Cloud computing is still an emerging technology within the AEC sector. The architecture industry has struggled to keep up with the trend. However, cloud platforms have become widespread in recent years with the increase of remote working. From increased access and data protection to better collaboration and rendering time, the cloud has revolutionized how architects work [14].

In a field that evolves at the speed of innovation, architects must stay ahead of the curve. Cloud computing has the potential to become the basis for plenty of potentially industry-changing developments. The universal call to utilize cloud technologies to their full potential reverberates throughout the architecture and design community. Architects seeking to embark on this transformative journey must elevate their design workflows, amplify collaboration, and embrace the future of architecture with confidence. Cloud-based computing for architects is here to stay. More information about cloud computing in architecture can be found in the books in [15–18] and the following related journal: *Journal of Cloud Computing.*

References

1. B. Libby, Designing in a cloud: how architects are using cloud-based programs to deliver the best projects (2014). https://www.architectmagazine.com/practice/best-practices/designing-in-a-cloud-how-architects-are-using-cloud-based-programs-to-deliver-the-best-projects_o
2. Cloud architect role: What is a cloud architect, Unknown Source
3. M.N.O. Sadiku, U.C. Chukwu, J.O. Sadiku, Cloud computing in architecture. Innov. Mult-Discip. J. Appl. Technol. **3**(1), 16–27 (2025)
4. S. Ezell, B. Swanson, How cloud computing enables modern manufacturing (2017). https://itif.org/publications/2017/06/22/how-cloud-computing-enables-modernmanufacturing
5. "Cloud computing applications in agriculture. https://www.eescorporation.com/cloud-computing-applications-in-agriculture/
6. Cloud manufacturing, *Wikipedia*, the free encyclopedia. https://en.wikipedia.org/wiki/€loud_manufacturing
7. J.S. Saini et al., [Retracted] cloud computing: legal issues and provision. Secur. Commun. Netw. 2022. https://www.hidawi.com/jo rnals/scn/2()22/2288961/
8. D. Rekko, Cloud-based computing for architects is already here (2024). https://www.designair.io/blogs/read-more/cloud-based-computing-for-architects-is-already-here
9. A. Centino, How architecture offices can leverage cloud computers (2023). https://vagon.io/blog/how-architecture-offices-can-leverage-cloud
10. R. Brady, How architects can supercharge their project workflow using the cloud. https://archdaily.com/blog/practice/tools/architects-cloud-computing/
11. What is BIM, https://bimmda.com/en/what-is-bim
12. T.H. Beach, Cloud computing for the architecture, engineering & construction sector: requirements, prototype & experience. J. Cloud Comput. **2**(8) (2013)
13. You're looking to advance your career in architecture. How can cloud computing help you get there? https://www.linkedin.com/advice/3/youre-looking-advance-your-career-architecture-sq8wc
14. A. Murtagh, Cloud computing for architects: how to increase efficiency (2022). https://www.rock.co.uk/insights/cloud-computing-architects-how-increase-efficiency/
15. M.N.O. Sadiku, *Cloud Computing and Its Applications* (Lambert Academic Publishing, Moldova, Europe, 2024)
16. B. Furht, A. Escalante, *Handbook of Cloud Computing* (Springer Science Business Media, New York, 2010)
17. T. Erl, E. Monoy, *Cloud Computing: Concepts, Technology, Security, and Architecture.* Pearson, 2nd edition (2023)
18. G. Reese, *Cloud Application Architectures: Building Applications and Infrastructure in the Cloud* (O'Reilly Media, 2009)

Chapter 10
Blockchain in Architecture

Blockchain facilitates democratic decision-making by enabling collective rule-setting and building in checks and balances.—Lan van Wassenaer.

10.1 Introduction

Blockchain is a technology that allows the recording of information in a way that it is difficult or practically impossible to alter, hack, or cheat. It is regarded as a promising general-purpose technology (GPT) in that it has the potential to affect an entire economy, impacting economic growth and transforming both everyday life and the ways in which we conduct business. Given its unique characteristics, such as immutability, transparency, decentralization, and distribution, blockchain is recognized by many as a new form of GPT. The symbol of a blockchain is depicted in Fig. 10.1 [1].

Construction is one of the largest industries in the world, creating infrastructure which is the backbone of productivity and economic growth. The architecture, engineering, and construction (AEC) industry has entered a period of major disruption caused by a host of emerging digital technologies such as artificial intelligence (AI), the Internet of things (IoT), virtual reality (VR), augmented reality (AR), building information modelling (BIM), 3D printing, robotics, drones, big data, and others. Lately, another innovative digital technology, blockchain, has appeared, promising to change the way people do transactions, keep records, validate data, and much more. Blockchain is a new general-purpose technology that can transform both our everyday lives and the ways we do business [2].

Blockchain is a decentralized system that uses a peer-to-peer network to create a digital ledger of transactions. It is a decentralized ledger with a growing list of transaction records. It is a promising general-purpose technology that can affect an entire economy and transform both our everyday lives and the ways we do business. It can simplify and secure transactions among parties and record keeping in general. Architects can utilize blockchain technology to enhance the transparency, security,

M. N. O. Sadiku, *Emerging Technologies in Architecture*,
https://doi.org/10.1007/978-3-031-99509-5_10

Fig. 10.1 The symbol of blockchain [1]

and efficiency of their design and construction process by creating a decentralized and immutable record of project data [3].

The aim of this chapter is to explore the use of blockchain technology in architecture. It begins with explaining the concept of blockchain. It describes blockchain in architecture and provides some of its applications. It highlights the benefits and challenges of blockchain in architecture. The last section concludes with comments.

10.2 What is Blockchain?

Blockchain, a type of distributed digital ledger technology (DLT), is a relatively new and exciting way of recording transactions in the digital age. It is a decentralized and distributed digital ledger technology that securely records and verifies transactions across multiple computers or nodes in a network. Basically, it is a chain of blocks in which each block contains a list of transactions. The blockchain technology was created as the foundational basis for Bitcoin–a digital currency in which secure peer-to-peer transactions occur over the Internet. It is expected that the spending on blockchain solutions worldwide would grow from 4.5 billion USD (2020) to an estimated value of 19 billion USD by 2024 [4].

Originally developed as the accounting method for the virtual currency Bitcoin, Blockchains are appearing in a variety of commercial applications today. Blockchain technology is a type of distributed digital ledger that uses encryption to make entries permanent and tamper-proof and can be programmed to record financial transactions. It is used for secure transfer of money, assets, and information via a computer network such as the Internet without requiring a third-party intermediary. It is now being adopted across financial and non-financial sectors. As a catalyst for change, the

Blockchain technology is going to change the business world and financial matters in major ways.

The first Blockchain was conceived in 2008 by an anonymous person or group known as Satoshi Nakamoto, who published a white paper introducing the concept of a peer-to-peer electronic cash system he called Bitcoin [5, 6]. Bitcoin and Ethereum are the first two mainstream blockchains. Other modern blockchains include Namecoin, Peercoin, Ether, and Litecoin. Figure 10.2 shows different components of blockchain [7].

Blockchain combines existing technologies such as distributed digital ledgers, encryption, immutable records management, asset tokenization and decentralized governance to capture and record information that participants in a network need to interact and transact. As illustrated in Fig. 10.3, a complete blockchain incorporates all the following five elements [8]:

Fig. 10.2 Different components of blockchain [7]

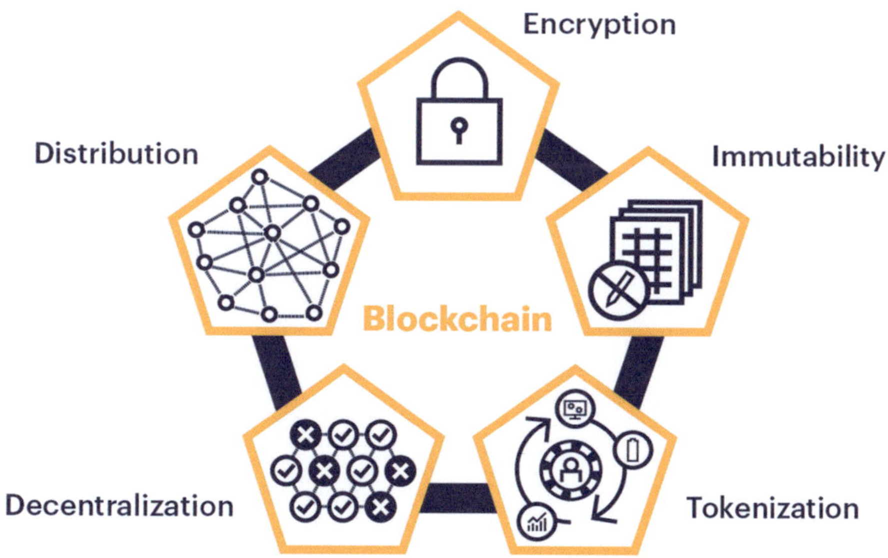

Fig. 10.3 Five key elements of Blockchain [8]

- *Distribution*: Digital assets are distributed, not copied or transferred. A protocol establishes a set of rules in the form of distributed mathematical computations that ensures the integrity of the data exchanged among a large number of computing devises without going though a trusted third party. A centralized architecture presents several issues including a single point of failure and problems of scalability.
- *Encryption*: BC uses technologies such as public and private keys to record data securely and semi-anonymously. Completed transactions are cryptographically signed, time-stamped, and sequentially added to the ledger.
- *Immutability*: The blockchain was designed so these transactions are immutable, i.e. they cannot be deleted. No entity can modify the transaction records. Thus, Blockchains are secure and meddle-free by design. Data can be distributed, but not copied.
- *Tokenization*: Value is exchanged in the form of tokens, which can represent a wide variety of asset types, including monetary assets, units of data or user identities.
- *Decentralization*: No single entity controls a majority of the nodes or dictates the rules. A consensus mechanism verifies and approves transactions, eliminating the need for a central intermediary to govern the network.

Bitcoin and its underlying blockchain technology increasingly impact all facets of society. Bitcoin's status as digital gold is merely the tip of this technology. Figure 10.4 shows Bitcoin [9], while Fig. 10.5 shows how blockchain works [10]. Although blockchain technology will for all time be associated with Bitcoin due to their common genesis, it has broader applications. Cryptocurrency will increasingly become a factor in family law issues as well.

A blockchain is a tamper-proof, distributed database that stores blocks of information for cryptographically bound transactions via peer-to-peer networks. At the heart

Fig. 10.4 Bitcoin [9]

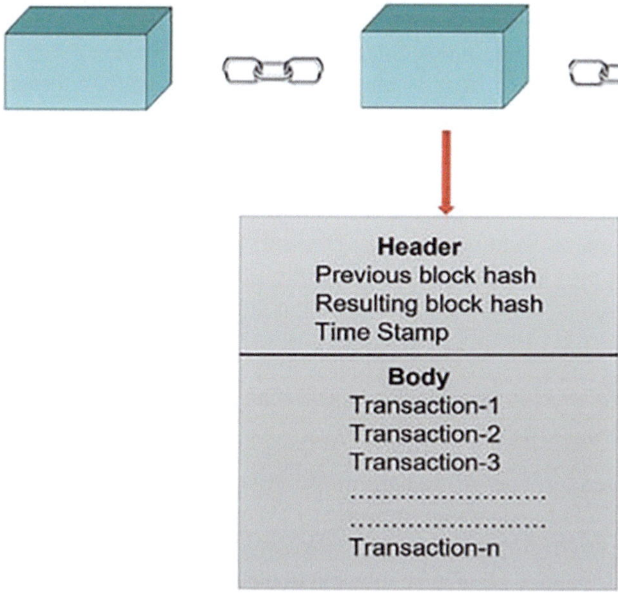

Fig. 10.5 How blockchain works [10]

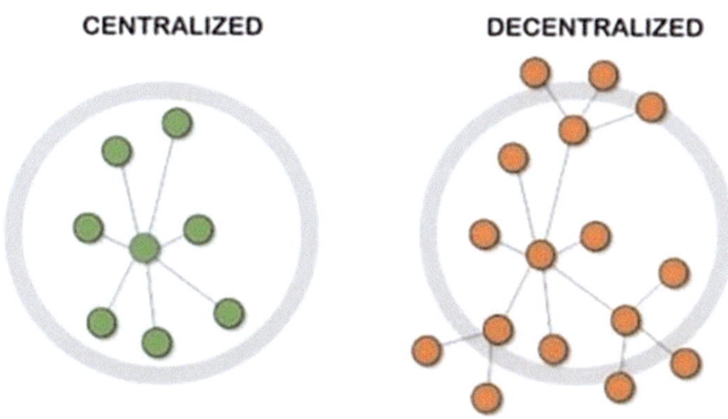

Fig. 10.6 The decentralized property of blockchain [10]

of blockchain's functionality is cryptographic hashing. Each block in a blockchain contains a cryptographic hash of the previous block, creating an immutable chain of blocks. If anyone attempts to tamper with the data in a block, it would alter the block's hash. This would disrupt the entire chain, making it virtually impossible to manipulate. The security feature ensures data integrity and prevents unauthorized changes [11].

In a nutshell, blockchain technology involves three basic concepts [12]: (1) It is a system for recording a series of data items (such as transactions between parties); (2) It uses cryptography to make it difficult to tamper with past entries; (3) It has an agreed process for storing copies of the ledger and adding new entries (also called a consensus protocol).

Blockchain is a novel decentralized infrastructure and distributed computing paradigm that uses a chained data structure for verification, storage, and distributed consensus algorithms to generate and update data. Decentralization is a key feature of blockchain technology, which refers to the distribution of power and decision-making across a network of nodes or participants rather than being controlled by a central authority or system. It provides robustness while eliminating many-to-one traffic flows to avoid delays and single points of failure. 10. 6 shows the decentralized property of blockchain [10]. The advantages of decentralized property of blockchain network include the following [10]:

- The decentralized property of the blockchain makes it less prone to failure and more expensive for hackers to attack the network.
- There is no third-party involvement; therefore, there is no added risk.
- Every change made in the network is traceable and concrete.
- Users maintain full autonomy of their properties and are not dependent on third parties to maintain and manage their assets.
- It provides enhanced security.

10.3 Blockchain in Architecture

Blockchain technology can be used in architecture to improve the design process, increase efficiency, and enhance transparency. It has the potential of addressing common problems of the AEC industry, while it can be adaptable to the construction/architecture industry structure and the way it is practiced. Blockchain's immutability ensures that once data is recorded, it cannot be altered or tampered with. This feature can be beneficial for maintaining a secure and unalterable record of design iterations, decisions, and project milestones, providing a transparent audit trail throughout the design process. Blockchain technology can enhance the architectural design process by providing immutable records, smart contracts, improved supply chain management, and decentralized collaboration. It promotes transparency, efficiency, and better project outcomes.

10.4 Applications of Blockchain in Architecture

In this section we identify several key areas of the civil, architectural, and construction industry where blockchain can be applied. Blockchain technology has many uses in architecture, including data management, collaboration, and contract creation.

We identify five relevant areas of benefit, namely: (1) supply chain management, (2) building information modelling (BIM), (3) contract management, (4) project management, (5) smart cities, (6) real estate. These common applications are explained as follows [2]:

- *Supply Chain Management*: The construction industry is characterized by fragmentation in processes, operations, and services. One of its major problems is the disconnect between construction and design, due to the lack of trustworthy and open information across the supply chain. Blockchain has the potential to adverse these issues using open and transparent transactions. It can improve payment settlements, compliance management and material planning, while smart contracts can be implemented to automatically purchase, track, and verify items in the supply chain, in real-time.
- *Building Information Modelling*: BIM is a digital representation of physical and functional characteristics of a facility. It is known as the cutting-edge technology for use in AEC practices. It is a crucial part of promoting the effectiveness of project collaboration throughout the whole building lifecycle. Blockchain can be used to provide live and trustworthy information for BIM, by information sharing among present and future information owners. It can help enhance the benefits of BIM by allowing architects and engineers to design on the same BIM model with clear ownership, while design and construction decisions can be recorded on the blockchain for future analysis and liability. With BIM-blockchain integration, the AEC industry can avoid various issues and get several benefits. Figure 10.7 shows BIM in the architecture industry [13].
- *Contract Management and Smart Contracts*: A smart contract is a computer program that works based on an "if/then" principle. Smart contracts are executed automatically reducing the necessity of intermediaries and as a result time and money can be saved. Blockchain can be used to create project-specific contracts that are restricted to certified contractors. A blockchain-based contract can be created for a project, restricted to certified contractors, and voided if certain requirements are not met. Smart contracts on blockchain can automate contract execution based on pre-defined conditions, simplifying complex agreements between stakeholders and reducing administrative overhead. Decentralized smart contracts based on blockchain can address some limitations in an effective way. Smart contracts are used in smart buildings for repairs and service. Figure 10.8 shows smart contracts [14].
- *Project Management*: Construction project management (CPM) can potentially benefit from an agile and more decentralized approach based on blockchain, with high transparency. Blockchain has a credible potential in the construction industry, due to its exponential general use, the investments involved, and a number of start-up businesses contributing to Industry 4.0. A blockchain-based CPM platform implementing smart contract technologies facilitates the peer-to-peer collaboration between parties in the construction industry, leading to improved information flow, cost, and time reduction, and improvement in the quality of the services.

- *Smart Cities*: As urbanization is increasing rapidly, offering improved livability and a higher standard of living, the concept of "smart cities" is one of the main focus areas of many governments across the globe. Smart cities enable operational efficiency, maximize environmental sustainability efforts, and create new citizen services. Blockchain innovation can be utilized to make smarter cities. Blockchain-based solutions can be utilized to enhance our cities and provide for better economic development and livability, by offering enhanced security, immutability, resilience and transparency. Blockchain and smart contracts can securely and flexibly manage building access privileges for both short-term visitors and long-term occupants, taking into consideration the risk associated with accessing a space in the building, in an efficient, decentralized way. For example, the Smart Dubai initiative aims to use blockchain technology to transform the city into a smart and sustainable hub, by streamlining the processes and transactions of various sectors, including architecture and construction. As another example, cryptocurrency millionaire Jeffrey Berns has revealed plans to develop a large parcel of Nevada's desert into a smart city powered by blockchain technology. He intends to use the smart city as a new kind of business and residential community, with a decentralized blockchain infrastructure underlying all interaction. Such a planned smart city powered by blockchain is shown Fig. 10.9 [15]. A potential major advantage of this city model is greater control of privacy of personal information.
- *Real Estate*: Real estate is known as one of the most important sectors of the economy, playing a crucial role in the lives of people across the world. Real estate investments provide better returns than the stock market without as much volatility, providing also tax benefits in many cases. Real estate is still a "pen and pencil" business, relying on archaic methods for keeping records and doing transactions. Figure 10.10 presents the real estate ecosystem [2] today and depicts the number of different parties involved and the relevant interactions. This traditional model has several drawbacks and limitations. There are a lot of intermediaries that increase the cost and reduce the transactions' speed. Real estate is entering the blockchain era and it can benefit from the numerous advantages that the technology can offer. With Blockchain, all real estate ownership and transaction records can be stored securely as tamper-proof digital records on the blockchain, in a decentralized way.

10.5 Benefits

Blockchain technology can offer several benefits for the design process in architecture. Blockchain's decentralized and transparent nature can help improve the efficiency and security of the design process and can also help with collaboration and communication. Blockchain technology has the potential to enhance various aspects of the design process for architects by providing transparency, security, and efficiency.

The five greatest advantages of blockchain technology in industries are higher transparency, higher security, higher traceability, higher efficiency and speed, and lower costs. Other benefits of blockchain in architecture include the following [16, 17]:

- *Decentralization*: Architects often deal with large volumes of data, including design plans, specifications, and project documentation. Blockchain's decentralized nature allows for secure storage and sharing of data without reliance on a central authority. Centralization can undermine the technology's purpose as a shared ledger. Too much power can be placed in the control of a single entity or a few special "players." When a single entity takes too much power then it may no longer operate for the benefit of the society. Fully decentralized distributed systems can mitigate risk and prevent attacks while centralized systems are more prone to them.
- *Increased Transparency*: BC can enhance the transparency and trust of the design process, as it provides a secure and immutable record of the project history, decisions, and changes. It can increase the efficiency of processes within the construction industry and eliminate current issues related to trust, verification and transparency. It can create a secure and immutable record of project history, decisions, and changes. All project data is readily accessible and verifiable by all parties involved, reducing the risk of fraud or manipulation.
- *Enhanced Efficiency*: BC can increase the efficiency and accuracy of data management, as it eliminates the need for manual entry, duplication, and reconciliation of information. Streamlined workflows and automated processes can significantly reduce administrative burdens and accelerate project timelines. The AEC industry needs to complete the task much faster than possible with a single computer. It is essential to meet this need by adopting distributed ledger technologies (DLT).
- *Smart Contracts*: Smart contracts, self-executing contracts with the terms of the agreement directly written into code, can automate and enforce various aspects of project management. This includes automatically triggering payments, managing project milestones, and ensuring that contractual obligations are met, streamlining administrative processes.
- *Collaboration*: Blockchain technology can enhance the architectural design process by providing immutable records, smart contracts, improved supply chain management, and decentralized collaboration. Blockchain can facilitate the collaboration and communication of the design team, as it enables real-time sharing and verification of data, documents, and contracts. It allows design teams to share and verify data, documents, and contracts in real time. A shared blockchain platform can enable real-time data sharing between architects, engineers, contractors, and clients, fostering better communication and coordination.
- *Secure Storage*: Conventional cloud storage and computing in building information modeling (BIM) are ineffective in providing trusted networks. Nowadays, the need for distributed storage and computing has been highlighted by the AEC industry. Blockchain's decentralized nature allows for secure storage of large amounts of data, such as design plans and specifications. Traditional

cloud computing environment has a data center that contains many computers. One major issue in such centralized storage and computing concerns the risk of operating failure. Trusted networks are urgently needed for data storage and computing. Applying distributed storage and computing can essentially reduce the risk of failure and provide trusted networks.

- *Clash detection*: Digital tools based on BIM provide automatic quantity take-off and clash detection. Such BIM technology has been widely applied in clash management that includes clash detection and resolution during design coordination of buildings and infrastructures. Generally, BIM coordinators conduct clash detection by merging architectural, structural, and mechanical, electrical, and plumbing (MEP) models in an integrated BIM platform, launching automatic clash detection algorithms, and then generating clash detection reports.
- *Improved Accountability*: Each stakeholder's actions and decisions are clearly recorded on the blockchain, enhancing accountability and responsibility. Blockchain introduces a new level of responsibility in construction. Everyone involved gets access to the same unchangeable data, ensuring each party is held accountable for their actions. It leads to a more honest and responsible way of implementing construction projects.
- *Tracking Design Changes*: Recording every design modification on the blockchain ensures a complete, verifiable history of project evolution, eliminating discrepancies and facilitating easier conflict resolution.
- *Material Traceability*: By recording the origin and provenance of building materials on the blockchain, architects can verify sustainability claims and ensure materials meet quality standards.
- *Reduced Costs*: By minimizing errors and disputes, blockchain can potentially lead to cost savings throughout the construction process

10.6 Challenges

Blockchain technology poses some challenges for the design process in architecture. Like the Internet in its first years, blockchain is difficult to predict or even understand well. Information exchange in BIM is critical yet complex due to the multi-party collaboration nature of a construction project. Other challenges include the following [16, 17]:

- *Resistance to Change*: The construction industry has traditionally been very reluctant to change and rather slow in adopting new digitization technologies. It can be difficult to persuade stakeholders to invest in this technology, particularly if they are used to conventional approaches. It is essential to show the long-term advantages and return on investment in order to get over change resistance.
- *Data Privacy and Security*: In a world where massive data is generated daily, data privacy has become a major concern. Blockchain goes a long way in addressing this issue. Given that the construction business deals with sensitive information and intellectual property, data privacy and security are major issues. Sensitive

data is used in construction projects, such as confidential financial information and private designs. Blockchain technology enhances security through encryption and decentralization, but businesses must still take into account the privacy implications and make sure they are in compliance with all applicable laws.

- *Standardization*: For best practices to be established across the sector and to encourage wider adoption, collaboration, and standardization efforts are essential. Standardization promotes widespread usage of blockchain technology in the construction industry and facilitates frictionless data interchange while increasing transparency. To create industry-wide standards, protocols, and governance models, coordination is required between architectural businesses, technology providers, and regulatory organizations.
- *Scalability*: One of the main challenges is the scalability and performance of the system, as it requires a large amount of computing power and storage capacity to process and store the data.
- *Interoperability*: Another challenge is the interoperability and compatibility of the system, as it requires a common standard and protocol to exchange and integrate data across different platforms and applications.
- *Regulation*: Another challenge is the regulation and governance of the system, as it requires a clear and consistent legal framework and ethical guidelines to protect the rights and responsibilities of the stakeholders.
- *Cyberattacks*: It was observed that the construction industry was the third target of ransomware attacks in North America in 2020. BIM-collaborative design is one of the most vulnerable phases to cyberattacks. Once a malicious insider or an outside attacker performs a successful breach on a single centralized database, the process of recovery can be complex and uncertain.

10.7 Conclusion

Blockchain technology is still in its early stages of development and adoption in the design process in architecture. It has the potential to revolutionize the way architects design, communicate, and collaborate. It will play a significant role in architecture industry in the future. In the future, it could become ubiquitous in the exchange of physical and digital goods, record keeping, information, and online platforms. Although blockchain technology is new and there are certainly several early challenges to tackle, it has great potential to become an extremely positive force of change in the architecture industry.

To leverage blockchain technology in architectural design, architects must start by familiarizing themselves with its fundamentals and potential impacts on the industry. By leveraging the power of blockchain technology, architects can improve the quality, efficiency, and innovation of their design solutions, as well as the trust and engagement of their clients and partners. Although nobody can predict the degree to which it will affect the economy, our lives and every single sector including architecture, most experts agree that it has the potential to play a significant role in the future.

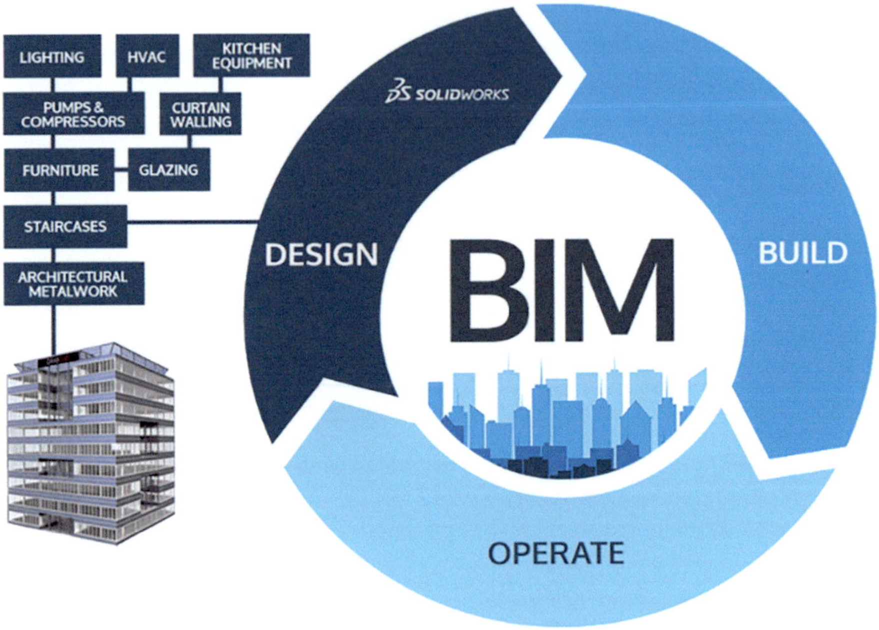

Fig. 10.7 BIM in the architecture industry [13]

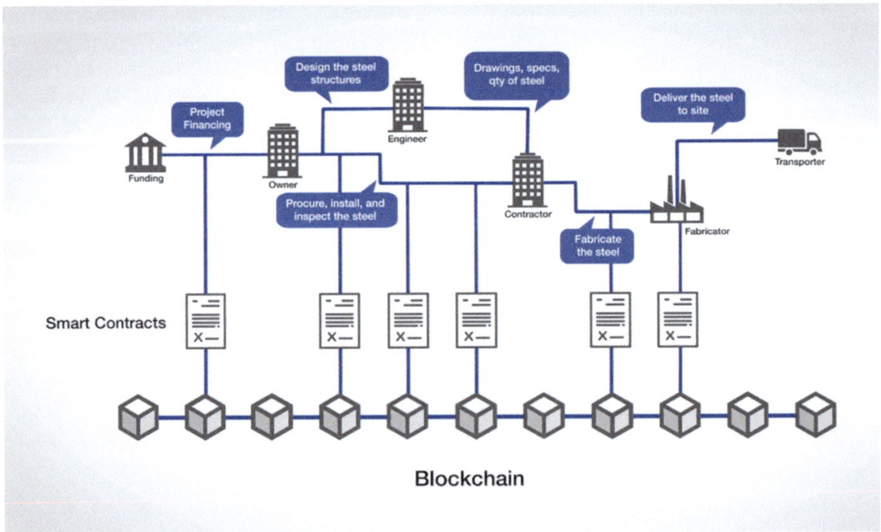

Fig. 10.8 Smart contracts [14]

Fig. 10.9 A planned smart city powered by blockchain [15]

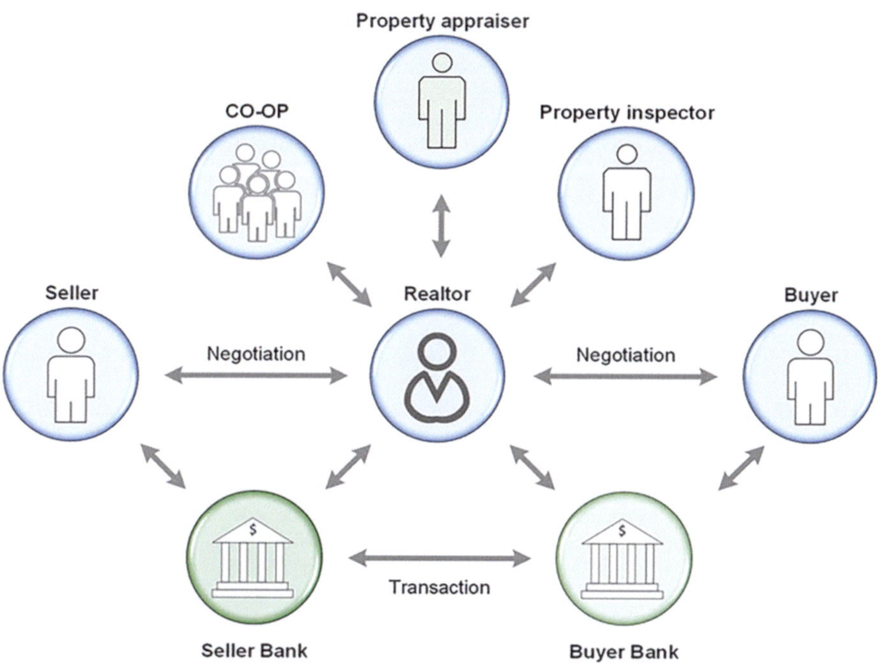

Fig. 10.10 The real estate ecosystem [2]

More information on the integration of blockchain technology into the architecture industry is available from the books in [18–22] and in the following related journals:

- *IEEE Blockchain*
- *Journal of Building Engineering*
- *Journal of Construction Engineering and Management*

References

1. Blockchain meets the oil & gas industry (2018). https://executiveacademy.at/en/news/detail/blockchain-meets-the-oil-gas-industry/
2. V. Plevris, N.D. Lagaros, A. Zeytinci, Blockchain in civil engineering, architecture and construction industry: State of the art, evolution, challenges and opportunities. https://www.frontiersin.org/journals/built-environment/articles/https://doi.org/10.3389/fbuil.2022.840303/full
3. M.N.O. Sadiku, P.A. Adekunte, J.O. Sadiku, Blockchain in architecture. Int. J. Trend Sci. Res. Dev. **9**(1), 367–376 (2025)
4. C.M.M. Kotteti, M.N.O. Sadiku, Blockchain technology. Int. J. Trend Res. Dev. **10**(3), 274–276 (2023)
5. Blockchain, *Wikipedia,* the free encyclopedia. https://en.wikipedia.org/wiki/Blockchain
6. S. Nakamoto, Bitcoin: a peer-to-peer electronic cash system. https://bitcoin.org/bitcoin.pdf
7. The beginning of a new era in technology: Blockchain traceability. https://www.visiott.com/blog/blockchain-traceability/#:~:text=The%20Beginning%20of%20a%20New,money%20without%20a%20central%20bank
8. The CIO's guide to blockchain. https://www.gartner.com/smarterwithgartner/the-cios-guide-to-blockchain#:~:text=True%20blockchain%20has%20five%20elements,%2C%20immutability%2C%20tokenization%20and%20decentralization.
9. Blockchain and space exploration: Is decentralized data the future of space missions? (2024). https://medium.com/coinmonks/is-decentralized-data-the-future-of-space-missions-646173d1aeec
10. B.G. Celik, Y.S. Abraham, M. Attaran, Unlocking blockchain in construction: a systematic review of applications and barriers. Buildings **14**(6) (2024)
11. D. Singh, Blockchain in pharmaceutical supply chain: the next big frontier (2023). https://www.debutinfotech.com/blog/blockchain-in-pharmaceutical-supply-chain-the-next-big-frontier
12. D. Michels, Technology blockchain and telecoms. https://www.iicom.org/wp-content/uploads/22-26-blockchain.pdf
13. SOLIDWORKS & BIM in the architectural & construction industries. https://www.innova-systems.co.uk/solidworks-architectural-construction-industry-bim/
14. K.A. Anand, What is blockchain technology in construction? (2023). https://www.getpowerplay.in/resources/blogs/what-is-blockchain-technology-in-construction/
15. E. Gibson, Cryptocurrency millionaire plans blockchain smart city in Nevada desert (2018). https://www.dezeen.com/2018/11/06/blockchain-innovation-park-smart-city-nevada-jeffrey-berns/
16. How can blockchain technology improve the design process for architects? https://www.linkedin.com/advice/0/how-can-blockchain-technology-improve-design-process-epo6e#:~:text=Blockchain%20technology%20can%20enhance%20the,efficiency%2C%20and%20better%20project%20outcomes
17. T. Zhang, D.T. Doan, J. Kang, Application of building information modeling-blockchain integration in the architecture, engineering, and construction/facilities management industry: a review. J. Build. Eng. **77** (2023)

18. M.N.O. Sadiku, *Blockchain Technology and Its Applications* (Lambert Academic Publishing, Moldova, Europe, 2023)
19. T. Dounas, D. Lombardi (eds.), *Blockchain for Construction* (Springer, 2022)
20. F. Elghaish et al., *Blockchain of Things and Deep Learning Applications in Construction: Digital Construction Transformation* (Springer, 2023)
21. S. Aghili, *The Auditor's Guide to Blockchain Technology: Architecture, Use Cases, Security and Assurance* (CRC Press, Boca Raton, FL, 2022)
22. K. Werbach, *The Blockchain and the New Architecture of Trust* (The MIT Press, 2018)

Chapter 11
3D Printing in Architecture

3D Printing has the potential to revolutionize the way we make almost everything.—Barack Obama.

11.1 Introduction

A 3D printer has more manufacturing capacity than a traditional manufacturing machine. It is regarded as a disruptive technology that will change manufacturing. It has been used for decades in the automotive and aerospace industries. The 3D printer is also used by hobbyists, small businesses, creatives, manufacturers, architects, and most importantly contractors to instantly create a variety of products.

Architecture is both the process and product of planning, designing, and constructing physical structures. It is one of the industries that combine art and technology seamlessly. The amalgamation of art and technology enables architects to plan and design a wide variety of structures that keep onlookers mesmerized. 3D printing, also known as additive manufacturing, revolutionizes traditional architectural processes by layering materials to create three-dimensional structures from digital models.

Since the time of the Pharaohs, architectural models have served as physical representations to help sell a project, support fundraising efforts, and solve construction challenges. Traditionally, model-making is a manual craft that involves working with materials such as wood, ceramic, cardboard, or clay and is extremely time-consuming. Today, architecture studios and practices have access to a wider range of tools, including CAD software, Building Information Modeling (BIM) solutions, and 3D printers that can reduce labor needs and save time [1].

3D printing or additive manufacturing is the process of creating physical items by layering materials based on the requirements of a digital model. The technology has become increasingly popular in recent years, and it is now used in various industries. One area where 3D printing is making a significant impact is architecture. Manually curating physical architectural models can be tedious and time-consuming. 3D

M. N. O. Sadiku, *Emerging Technologies in Architecture*,
https://doi.org/10.1007/978-3-031-99509-5_11

printing is poised to revolutionize the architecture industry. There is no doubt that 3D printing is a disruptive technology set to change how we build our cities [2].

This chapter examines the use of 3D printing technology in the architecture industry. It begins with explaining what 3D printing is all about. It presents 3D printing in the architecture industry. It provides several applications of 3D printing in the architecture industry. It highlights the benefits of and challenges of 3D printing in architecture. It concludes with some comments.

11.2 What is 3D Printing?

3D printing (also known as additive manufacturing (AM) or rapid prototyping (RP)) was invented in the early 1980s by Charles Hull, who is regarded as the father of 3D printing. Since then it has been used in manufacturing, automotive, electronics, aviation, aerospace, aeronautics, engineering, architecture, pharmaceutics, consumer products, education, entertainment, medicine, space missions, the military, chemical industry, maritime industry, printing industry, and jewelry industry [3].

A 3D printer works by "printing" objects. Instead of using ink, it uses more substantive materials–plastics, metal, rubber, and the like. It scans an object–or takes an existing scan of an object–and slices it into layers, which can then convert into a physical object. Layer by layer, the 3D printer can replicate images created in CAD programs. In other words, 3D printing instructs a computer to apply layer upon layer of a specific material (such as plastic or metal) until the final product is built. This is distinct from conventional manufacturing methods, which often rely on removal (by cutting, drilling, chopping, grinding, forging, etc.) instead of addition. Models can be multi-colored to highlight important features, such as tumors, cavities, and vascular tracks. 3DP technology can build a 3D object in almost any shape imaginable as defined in a computer-aided design (CAD) file. It is additive technology as distinct from traditional manufacturing techniques, which are subtractive processes in which the material is removed by cutting or drilling [4].

3D printing has started breaking through into the mainstream in recent years, with some models becoming affordable enough for home use. Many industries and professions around the world now use 3D printing. It plays a key role in making companies more competitive. The gap between industry and graduating students can be bridged by including the same cutting-edge tools, such as 3D printing, professionals use every day into the curriculum. There are 3D printed homes, prosthetics, surgical devices, drones, hearing aids, and electric engine components. As shown in Fig. 11.1, 3D printing involves three steps [5]. A typical 3D printer is shown in Fig. 11.2 [6].

The architectural designing community has embraced 3D printing with open arms and harnessed its limitless potential. Architects have been using 3D printers for years. 3D printer manufacturers offer a wide range of resources covering printers, technologies, software, and troubleshooting to make you an expert in no time. To

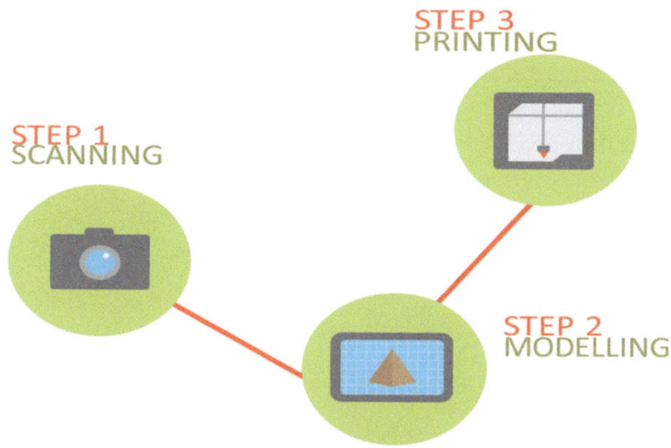

Fig. 11.1 3D printing involves three steps [5]

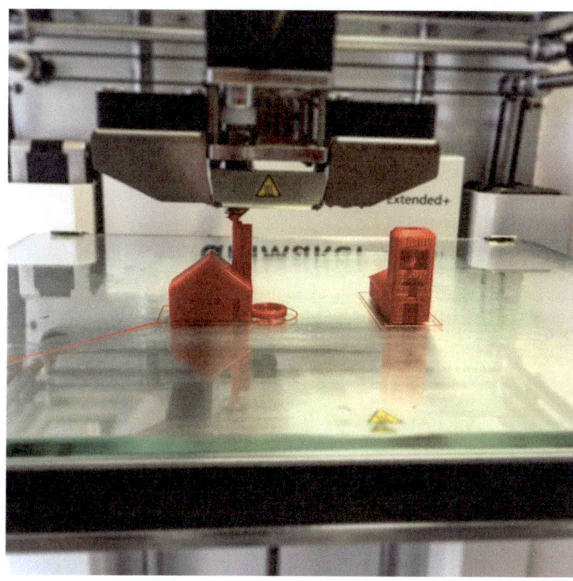

Fig. 11.2 A typical 3D printer [6]

fulfill the enormous demands of customers, companies are really pushing their limits hard to offer something out-of-the-box concept to their clients.

11.3 3D Printing in Architecture

Architectural models are conventionally assembled with a variety of materials and components. Modern CAD platforms have dedicated 3D printing modules to help architects convert a CAD design into a printable model. 3D printers help fuse these components into as few individual parts as possible. The size and geometry of different components within an architectural model are key considerations when preparing an architectural model for 3D printing.

How does 3D printing work in architecture? Architects must first create a digital model of the desired structure. Once the model is complete, it is sent to a 3D printer. Operating as a layer-by-layer additive manufacturing process, 3D printing uses digital models to create customized 3D objects with a remarkable level of precision and efficiency, saving time, generating zero waste, and reducing labor costs. The printer uses materials such as plastic, metal, or concrete. This process can take anywhere from a few hours to several days, depending on the size and complexity of the project. Post-processing differs depending on your specific 3D printing technology, but generally includes sanding, bonding, and painting models [1]. Figure 11.3 illustrates the 3D printing process in architecture [7], while Fig. 11.4 shows how 3D printing works in architecture [8].

Materials play a vital role in conveying the underlying concept of a design. Parts can be produced with various 3D printing materials, or individually painted with different colors. Some printers only print with their own proprietary materials, while others work with a wide range of materials. One of the most exciting developments in 3D printing is the use of more natural materials that allow for more environmentally conscious building techniques.

3D-printed homes have the potential to revolutionize construction and reduce the intensive carbon footprint of that industry. Key innovators in the industry include ICON, MOBBOT, and COBOD. They intend to expand into other markets to make their materials not only residentially available but also widely commercially available. If these innovators can show that their designs are safe, durable, and sustainable when compared to traditional homes, 3D-printing may become a viable and sustainable construction building method. Figure 11.5 shows a typical 3D printing building [9]. Figure 11.6 depicts the largest 3D-printed structure in Europe [10]. It is called Wave House and is used as cloud-computing data center in Heidelberg, Germany.

11.4 Applications of 3D Printing in Architecture

A rapidly growing number of companies all over the globe are choosing to incorporate additive manufacturing in their production processes. There are different applications of this advanced technology within the architecture industry. These include the following [11, 12]:

1. *Prototyping*: This is extremely important in architecture, and it is the stage that is being most heavily impacted by 3D printing technologies. Architects have quickly realized that 3D printing is ideal for prototyping and testing concept models. Due to the flexible nature of 3D printers, you can produce simple and complex building or structural models in hours instead of weeks or even months. Once you have your model you can begin discussions with clients, which in turn saves time and money, reduces the risk of costly errors, and speeds up the

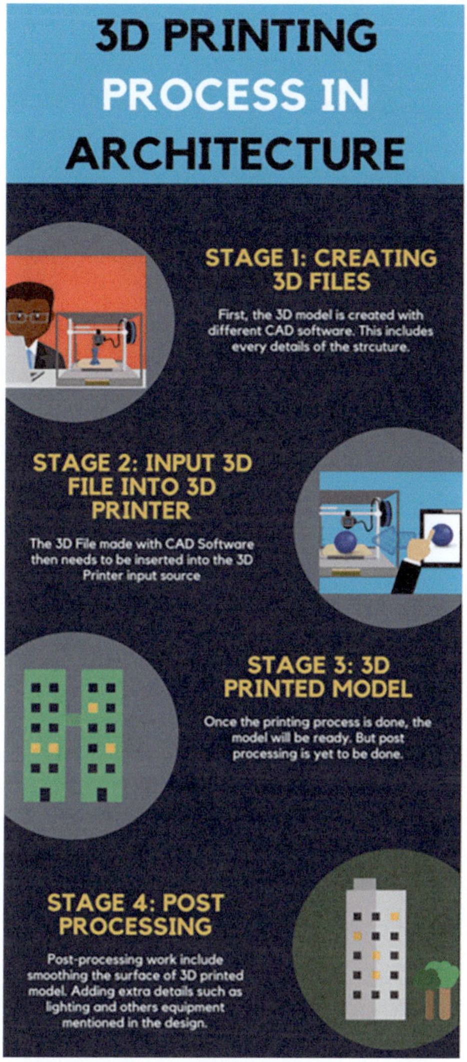

Fig. 11.3 3D printing process in architecture [7]

entire design-to-agreement process. The use of 3D rapid prototyping can reduce the manpower and time required, create detailed and accurate scale models, and effectively convey the unique concept of each design to the client.

2. *Digital 3D Modeling*: Previously, the creation of cardboard 3D models took a long time in addition to being labor-intensive. Today, 3D printing allows one to create high-quality 3-dimensional prototypes, significantly reducing time and effort. 3D modeling is a key step in the 3D printing workflow. Modeling can be used to design prototypes of a part, actual structures or scale models of a project. Some companies prefer to use several different software programs to achieve the

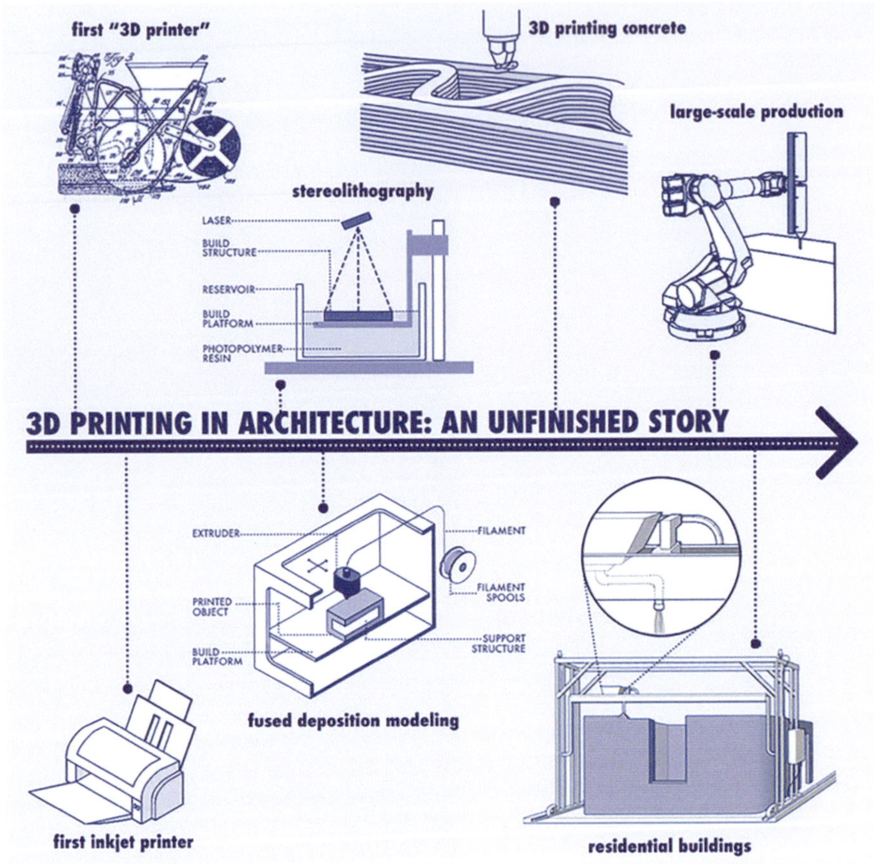

Fig. 11.4 How 3D printing works in architecture [8]

best results. Once the digital file is rendered, the designers typically delegate to the engineers to start 3D printing the prototype.

3. *Urban Infrastructure*: 3D printing is revolutionizing urban planning in many ways. Architects have the ability to 3D print a model of an entire town or city. This is something that is achievable within hours with the right equipment and print materials. Urban infrastructure 3D printing is often a collaboration between a 3D printing architecture firm and the city or government. The structures usually range from 3D printed bridges to decorative building facades.
4. *Developing Concept Models*: The architects have to finetune the concept models regularly to meet the precise needs and instructions of a client. They have to make changes to the concept model regularly to showcase all important aspects of the model. 3D printing can create highly detailed models that are accurate and can be used to showcase ideas to clients. Today, many architecture firms leverage 3D printing technologies to reduce the time and efforts required to convert a concept

Fig. 11.5 A typical 3D printing building [9]

Fig. 11.6 The largest 3D-printed structure in Europe [10]

model into a physical model. The experts can now showcase their ideas and impress their clients with tangible models that take into account precise building or construction site information.

5. *Pitching to Clients*: In the highly competitive architecture industry, designers need to express their architectural ideas vividly to win more projects. Clients want to visualize proposed building components or parts to make informed choices. So, competitive architects pitch to their clients using 3D printed models because these are highly detailed, and they provide a clearer picture of what the final project will look like. Physical 3D building models make it easier for designers to engage their clients in a meaningful way. Comprehensive site plan models that impressively showcase structural components help increase the chances of a project's approval by city authorities as well as investors.
6. *Construction Automation*: The leading architecture firms and professionals deploy autonomous machines and custom robotic solutions to implement automation in construction. They can automate most aspects of a construction project by combining custom robotic solutions, artificial intelligence (AI) algorithms, and 3D printing technologies. 3D printers will reduce overall construction time and cost by producing internal walls, fixtures, and floorboards simultaneously.
7. *Building Components*: 3D printing can be used to create full-scale building components. 3D-printed building components, such as facades, partitions, and structural elements, offer cost-effective solutions with unparalleled design flexibility. By leveraging parametric design tools and digital fabrication techniques, architects can customize building components to meet specific project requirements while reducing construction costs and lead times.

11.5 Benefits

Construction and architecture, which are often wasteful professions responsible for about 40% of greenhouse gas emissions, are the two key domains where 3D printing can be beneficial. 3D Printing is highly advantageous since it can create new products very quickly and precisely. Architects are increasingly using 3D printers to produce beautiful and highly detailed architectural models. They are saving time and money by utilizing 3D printing to create impressive and durable designs straight from CAD data. 3D printing can be used to produce architectural models of entire buildings. Other benefits of using 3D printing in architecture include the following [13–15]:

1. *Your Clients Can Visualize Better*: One of the main goals of 3D architecture models is to simplify communication between architects and to make it easier to showcase plans to clients and impress them with a tangible model of any structure. Being able to physically see something tangible creates a lasting impression. As illustrated in Fig. 11.7, a 3D printed object makes your design easy to visualize [16]. So a detailed scale model of the final project can be a valuable way of communicating your design idea to your colleagues or clients.

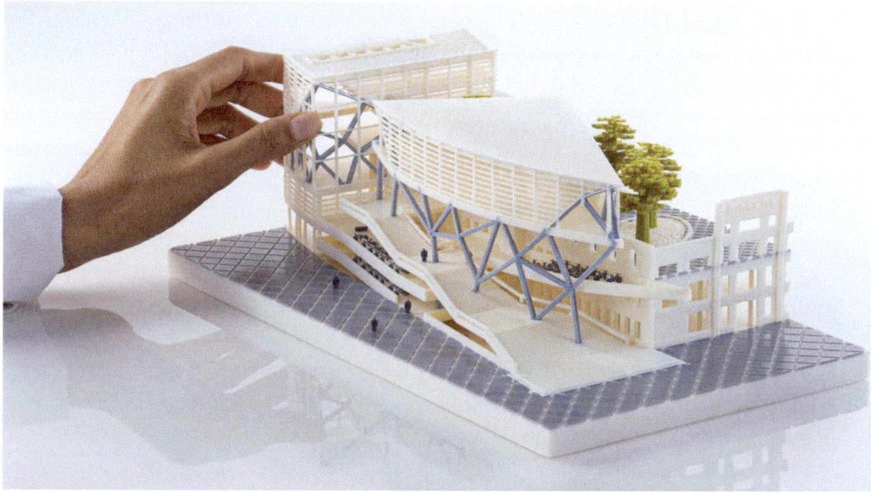

Fig. 11.7 3D printed object makes your design easy to visualize [16]

2. *Design Freedom*: The biggest advantage of applying 3D printing in architectural projects is the freedom of design, as shapes previously unimaginable have now become possible and even easy to manufacture. 3D printing stands poised to revolutionize the practice of architecture, offering architects unprecedented freedom to realize their creative vision and reimagine the built environment. With more design freedom comes the opportunity to create new complex geometries. Each client these days wants architects to design a building that stands out from the crowd. The architects have to explore ways to implement the exclusive models without increasing construction time and cost.
3. *Sustainability*: Additive manufacturing is more sustainable and environmentally friendly than other production methods. The sustainability is manifested in reduced material consumption, either considered reduced in the product itself, thanks to geometric optimization, or the various wastes associated with manufacturing like formworks. 3D printing holds immense promise for the creation of adaptive and sustainable structures that respond intelligently to environmental cues and user needs
4. *Customization*: The ability to offer highly customizable products has become key to stay competitive. Customization is hard to achieve with traditional techniques, where each mold needs to create a great number of products to justify the initial cost of investment.
5. *Collaboration*: 3D printing opens up numerous avenues for collaboration. A 3D modeling software using the cloud will allow you to optimize your team's collaboration and communication. Your work will be accessible to anybody, and you will all be able to work on the same model.

6. *Reduce Design Time*: In today's fast-paced construction industry, designers have to use the shortest time possible to develop 3D models. With 3D printers, you can produce complex building in hours rather than weeks or months. With a 3D printer you can make your design come to life while saving yourself the hours of creating a scale model. Time is money. So saving hours of work ultimately means saving expenses.
7. *Models In Great Quality*: Professional 3D printers now print architectural models with great quality and a great level of detail architects.
8. *High-Quality Prototypes*: 3D printers can produce high-quality architectural prototypes to clarify the details of exterior. The equipment may use different types of raw materials to deliver complex, color objects.
9. *Easily Re-Edit And Re-Print*: 3D printing allows you to be more flexible with your models since it is easy to re-edit and re-print your architectural 3D models. If the client requests a change, you can simply edit the file, and print the object.
10. *Saving Costs*: Architectural firms can acquire 3D printers for on-premises deployment, which is way cheaper than outsourcing to a provider. The equipment helps them enhance overall productivity and cut production costs.
11. *Saving Time*: 3D printers can operate overnight to save time. We can launch the printer in the night, and when we come back in the morning, we find the model done.
12. *Labor Force Reduction*: 3D printing technology involves low use of the labor force and saves a lot of time. While a set of labor is involved in one project, having a 3D printer allows them to distribute it in different other projects.
13. *Financially Beneficial*: As the architecture industry is a million-dollar industry, money is the game-changer. With in-house printers, it becomes an effortless modeling process. It is financially sustainable and impactful in the design method.

11.6 Challenges

3D printing does not allow us to print entire structures simultaneously, such as a building. The whole unit will be fabricated in parts at a manufacturing unit and then combined to create the final design at the construction site. Other challenges of 3D printing in architecture include the following [17]:

1. *High Cost*: 3D printing in architecture is still a new technology in the market. It requires a high initial cost of buying the necessary machinery and equipment. The additional cost of materials, experimentation, and maintenance poses one of the greatest barriers for anyone willing to try this technology.
2. *Lack of Known Materials*: Materials that can be printed and constructed in real time are essential for getting the best outcomes from 3D printing. Sophisticated printers can also work with metal, ceramic, and glass. Yet, the selection of materials that supports additive manufacturing is limited, restricting the widespread

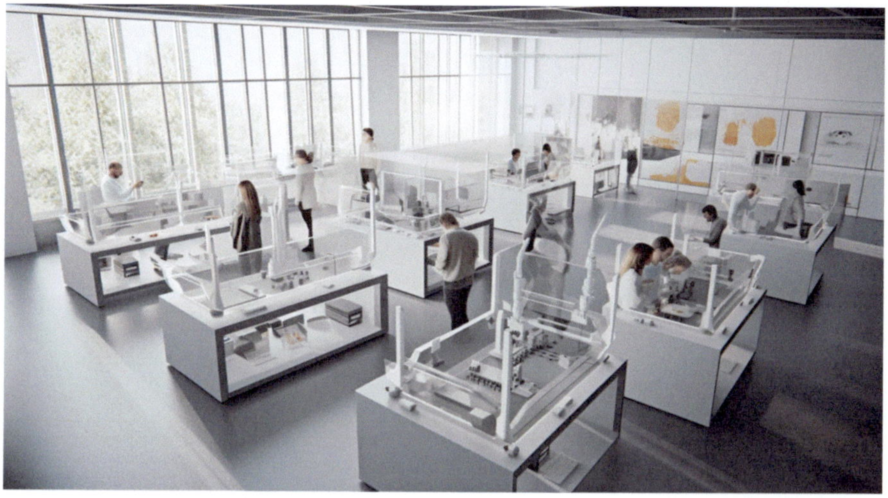

Fig. 11.8 A futuristic architecture class [18]

use of this technology. Not all construction materials are suitable for 3D printing, limiting the range of applications.

3. *Technical Limitations*: 3D printing faces technical challenges and limitations, such as size constraints, material compatibility, and post-processing. Certain architectural scales and complexities may be beyond current technological capabilities. Technological constraints such as printing speed, build volume, and material properties pose significant barriers to scalability and cost-effectiveness.
4. *Scale Limitations*: Larger projects require significant advancements in printing technology and materials. There is a growing need for larger machines capable of producing bigger parts or even entire structures.
5. *Regulatory Issues*: The innovative nature of 3D printing outpaces current building codes and standards, necessitating new frameworks for approval and implementation.

11.7 Conclusion

3D printing is revolutionizing the field of architecture, and it is poised to change the way we build our cities and towns in the future. This technology covers the way towards a bright future full of opportunities. Several universities are now offering courses in 3D printing for architects. Figure 11.8 shows a futuristic architecture class [18]. More information about 3D printing technology in the architecture industry can be found in the books [19–21].

References

1. Guide to 3D printing architectural models. https://formlabs.com/blog/3d-printing-architectural-models/
2. M.N.O. Sadiku, U.C. Chukwu, J.O. Sadiku, 3D printing in architecture. Innov. Mult-Discip. J. Appl. Technol. **2**(5), 136–145 (2024)
3. F.R. Ishengoma, T.A.B. Mtaho, 3D printing: developing countries perspectives computer engineering and applications. Int. J. Comput. Appl. **104**(11), 30–34 (2014)
4. M.N.O. Sadiku, S.M. Musa, O.S. Musa, 3D Printing in the chemical industry. Inven. J. Res. Technol. Eng. Manag. **2**(2), 24–26 (2018)
5. D. Pitukcharoen, 3D printing booklet for beginners. https://www.metmuseum.org/-/media/files/blogs/digital-media/3dprintingbookletforbeginners.pdf
6. 3D printing in architecture (2020). https://portella.com/blog/3d-printing-in-architecture/
7. S.R. Islam, Role of 3D printing in architecture. https://iamrafiqul.medium.com/3d-printing-in-architecture-eec154498852
8. Infographic: The evolution of 3D printing in architecture, since 1939. https://www.archdaily.com/1005043/infographic-the-evolution-of-3d-printing-in-architecture-since-1939
9. D. Gamsby, 3D printed architectural models are helping to build our future (2016). https://www.javelin-tech.com/blog/2016/05/3d-printed-architectural-models/
10. R. Kazmer, Group of architects create region's largest 3D-printed building in just 140 hours — take a look (2024). https://www.yahoo.com/tech/group-architects-create-regions-largest-080000044.html
11. M. Aysha, How are 3D printing technologies reshaping design and architecture? (2020). https://www.3dnatives.com/en/3d-printing-in-architecture-121120204/
12. A. Choudhry, 3D Printing in architecture: applications and innovations. https://www.aurum3d.com/blog/3d-printing-in-architecture-applications-and-innovations/
13. K. Ansari, What is 3D printing technology? How it is used in Architecture? (2021). https://archestudy.com/what-is-3d-printing-technology/
14. Top 4 benefits of 3D printing models for architects. https://i.materialise.com/blog/en/3d-printing-for-architects/
15. How architects are using 3D printing (2023). https://shop3duniverse.com/blogs/digital-fabrication-in-the-workplace/architects-using-3d-printing-today
16. N. Jaeger, The best 3D printers for architects of 2023 (2023). https://all3dp.com/1/the-best-3d-printers-for-architects/#google_vignette
17. The future of 3D printing in architecture: Opportunities and challenges. https://www.re-thinkingthefuture.com/architectural-community/a10290-the-future-of-3d-printing-in-architecture-opportunities-and-challenges/
18. 3D printing: Pioneering the future of architecture. https://insights.jonite.com/3d-printing-pioneering-the-future-of-architecture
19. C. Bañón, F. Raspall, *3D Printing Architecture: Workflows, Applications, and Trends* (Springer, Berlin/Heidelberg, Germany, 2021)
20. R. Rael, V. San Fratello, *Printing Architecture: Innovative Recipes For 3D Printing* (Princeton Architectural Press, 2011)
21. C. Banon, F. Raspall, *3D Printing Architecture: Workflows, Applications, and Trends*(Springer, 2021)

Chapter 12
Immersive Technologies in Architecture

VR and AR will eventually converge, and smart glasses will take over our digital interactions.—Carlos López.

12.1 Introduction

In the ever-evolving realm of architecture, the integration of cutting-edge technologies has revolutionized the way architects and designers conceptualize, plan, and communicate their ideas. Immersive technologies (ImT) have emerged as a groundbreaking tool that is transforming the field of architecture. They create highly interactive and engaging digital experiences, blurring the boundaries between the real and virtual.

Immersive technologies have the potential to impact the way we work, learn and interact with one another. They allow architects to step inside their designs, engage clients in immersive experiences, collaborate more effectively, and enhance overall project efficiency. Figure 12.1 shows a typical massive architectural design [1].

Immersive technologies, commonly used in gaming and entertainment, have become integral to architectural design, altering how spaces are envisioned. They create distinct experiences by merging the physical world with a digital or simulated reality. They are providing a new way for architects to design buildings and interior spaces, allowing for a more detailed and interactive experience for clients. They provide an improved means of communicating the experience of unbuilt designs to clients who cannot always envision a spatial experience from 2D representations. As immersive technologies have matured, they provide a powerful way to put that data to work.

An immersive technology is a dynamic emerging technology that challenges traditional notions of art and architecture. Immersive technologies have the potential to change how we communicate and learn by offering new opportunities for engagement, creativity, and problem-solving. Architects and designers use these technologies to create virtual walkthroughs, 3D visualizations, and interactive models of buildings and spaces. The use of immersive technologies in architecture is rapidly

M. N. O. Sadiku, *Emerging Technologies in Architecture*,
https://doi.org/10.1007/978-3-031-99509-5_12

Fig. 12.1 A typical massive architectural design [1]

becoming increasingly significant, and it is transforming the way architects design, conceptualize, communicate, construct, and collaborate. Immersive technologies like virtual reality (VR), augmented reality (AR), and mixed reality (MR) are used in architecture to help designers visualize and communicate their ideas. Architects use them to communicate their ideas to clients and colleagues [2].

In this chapter, we will explore how immersive technologies are reshaping our understanding of reality and virtual environments in architecture. The chapter begins with explaining what immersive technologies are. It describes immersive technologies in architecture and provides some of their applications. It highlights the benefits and challenges of immersive technologies in architecture. The last section concludes with comments.

12.2 What Are Immersive Technologies?

The first step in understanding how to use immersive technologies is to learn the differences between its various forms. In their simplest form, immersive technologies consist in adding virtual objects to the real world. There are four types of digital realities leading to different types of immersive technologies [3, 4]:

- *Augmented reality* (AR)—designed to add digital elements over real-world views with limited interaction.
- *Virtual reality* (VR)—immersive experiences helping to isolate users from the real world, usually via a headset device and headphones designed for such activities.

- *Mixed reality* (MR)—combining AR and VR elements so that digital objects can interact with the real world means businesses can design elements anchored within a real environment.
- *Extended reality* (XR)—covering all types of technologies that enhance our senses, including the three types previously mentioned.

These devices also enable new user interactions including spatially tracked 3D controllers, voice inputs, gaze tracking, and hand gesture controls.

Extended reality (XR) is the overarching term used to describe employing technology to blend real life and the digital world. It includes all the machine-human interfaces beyond the physical realm (reality) such as augmented reality (AR), mixed reality (MR), assisted reality (aR), and virtual reality (VR), as illustrated in Fig. 12.2 [5]. Figure 12.3 shows the XR spectrum [5]. Immersive technologies reside along a continuous scale ranging between the completely real and the completely virtual world. At one end, the real environment refers to the actual physical space, objects, and people that exist in the tangible world around us. At the other end, the virtual environment represents a completely computer-generated and immersive digital space, distinct from the physical reality. The space in the middle is called mixed reality, which is a blend of the real and virtual environments, where digital and physical elements coexist and interact in real time. A range of devices makes up XR, and these are used by consumers and in many industries for entertainment, safety, training, or productivity purposes.

1. *VIRTUAL REALITY*: Virtual reality (VR) is XR at its most extreme. It completely immerses the user in a digital world, often using a computer-generated environment with scenes and objects that appear to be real. The term "virtual reality" essentially means "near-reality." Virtual reality is the key technology for experiencing sensations of sight, hearing, and touch of the past, present, and future. VR is a fully immersive technology where users wear a head-mounted display and experience a simulated world of imagery and sounds. VR enables active learning. The terms, "virtual reality" and "cyberspace" are often used interchangeably. A cyberspace may be regarded as a networked virtual reality. A

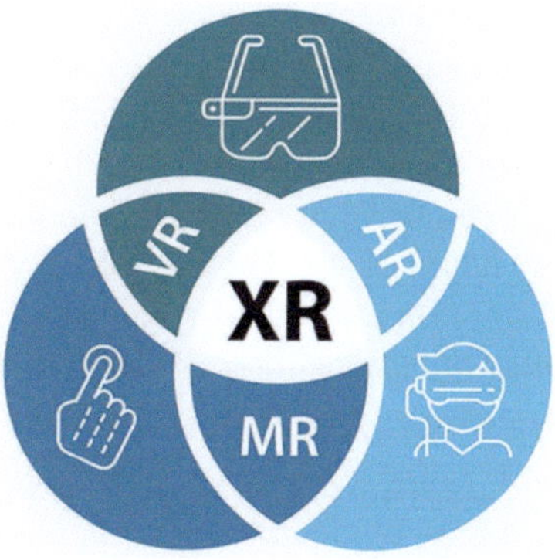

Fig. 12.2 Extended reality (XR) includes AR, MR, and VR [5]

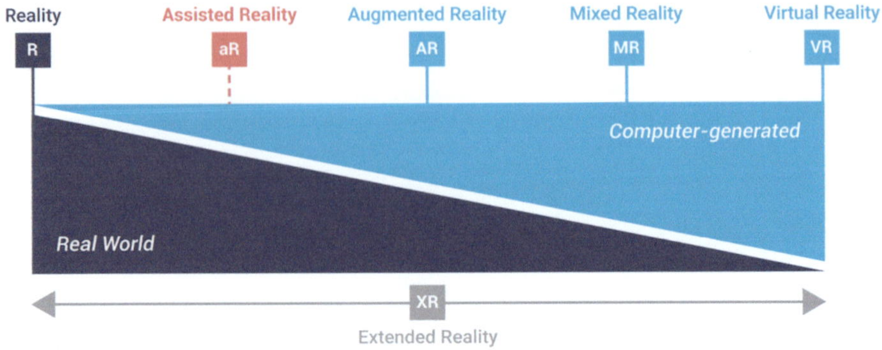

Fig. 12.3 The XR spectrum [6]

person using virtual reality can look around an artificial world, move around it, and interact with virtual features or items. This effect is commonly created by virtual reality headsets. Head-mounted displays immerse the user in a virtual environment. Virtual reality is a simulated experience that can be similar to or different from the real world. It is a computer-generated, 3D environment that completely immerses the senses of sight, sound, and touch. The complete immersion of the senses overwhelms users engrossing them in the action. Virtual reality technology includes multiple components divided into two main groups: hardware and software components [7].

- *Hardware Components*: The hardware components include a computer workstation, sensory displays, a tracking system, wearable devices, and input devices. Sensory displays are used to display the simulated virtual worlds to the user. The most common type is the head-mounted displays (HMDs), which is used in combination with tracking systems. Head-mounted displays are shown in Fig. 12.4 [8]. Users interact with the simulated environment through some wearable devices. VR depends on special responses such as raising hands, turning the head, or swinging the body. A wearable device is important in making these effects realistic. Special input devices are required to interact with the virtual world. These include the 3D mouse, the wired glove, motion controllers, and optical tracking sensors. These devices are used to stimulate our senses together to create the illusion of reality.
- *Software Components*: Besides the hardware, the underlying software plays an important role. It is responsible for the managing of I/O devices and time-critical applications. The software components are 3D modeling software, 2D graphics software, digital sound editing software, and VR simulation software. VR technology has been designed to ensure visual comfort and ergonomic usage.

2. *AUGMENTED REALITY*: Augmented reality (AR) is a technology that combines real-world environments with computer-generated generated information such

Fig. 12.4 Head-mounted display [8]

as images, text, videos, animations, and sound. It can record and analyze the environment in real-time. In augmented reality, the user typically experiences the real world through a device such as a smartphone, tablet, smart glasses, or head-mounted display. For example, AR allows consumers to visualize a product in more detail before they purchase it. This feature enhances consumer interaction and helps them never to repurchase the wrong item. The key objective of AR is to bring computer-generated objects into the real world and allows the user only to see them. In other words, we use AR to track the position and orientation of the user's head to enhance/augment their perception of the world. Augmented reality falls into two categories: 2D information overlays and 3D presentations, like those used with games. AR blends the virtual and real worlds by overlaying digital objects and information onto the users' view of the physical world.

To obtain a sufficiently accurate representation of reality, AR needs the following five components [9]:

- *Sensors*: AR needs suitable sensors in the environment and possibly on a user, including fine-grained geolocation and image recognition. These are activating elements that trigger the display of virtual information.
- *Image augmentation*: This requires techniques such as image processing and face recognition.
- *Head-mounted Display*: HMDs are used to view the augmented world where the virtual computer-generated information is properly aligned with the real world. Display technologies are of two types: video display and optical see-through display.
- *User Interface*: This includes technologies for input modalities that include gaze tracking, touch, and gesture. AR is a user interface technology in which

a camera-recorded view of the real world is augmented with computer-generated content such as graphics, animations, and 2D or 3D models.
- *Information infrastructure*: AR requires significant computing and communications infrastructure undergirding all these technologies. The infrastructure determines what real-world components to augment, with what, and when.

3. *MIXED REALITY*: Mixed reality (MR) is a term used to describe the merging of a real-world environment and a computer-generated one. Physical and virtual objects may co-exist in mixed reality environments and interact in real time. This is an extension of AR that allows real and virtual elements to interact in an environment. MR liberates us from screen-bound experiences by offering instinctual interactions with data in our living spaces and with our friends. Online explorers, in hundreds of millions around the world, have experienced mixed reality through their handheld devices. Mixed reality is a blend of physical and digital worlds, unlocking natural and intuitive 3D human, computer, and environmental interactions, as shown in Fig. 12.5 [10]. This new reality is based on advancements in computer vision, graphical processing, display technologies, input systems, and cloud computing. Mixed reality has been used in applications across fields including design, education, entertainment, military training, healthcare, product content management, and human-in-the-loop operation of robots [11].
4. *ASSISTED REALITY*: Like mixed reality, assisted reality (aR) is an extension of augmented reality, with a few notable differences to both. One of these differences is that aR is primarily hands-free through the wearing of a headset, whereas AR usually requires the holding of a device such as a mobile phone. While MR is a digital-first, real-world second reality, aR is a real-world first system. It combines software and a head-mounted display. It is best experienced using smart glasses or other wearable technology. The aR market is growing rapidly and promises to be the next great leap to boost workers' productivity. A worker wearing an aR device is shown in Fig. 12.6 [12].
5. *EXTENDED REALITY*: The term "extended reality" (XR) has recently gained favor as an umbrella term that encompasses all of AR, VR, and MR. The primary user inputs for XR devices are described as follows. Voice interfaces are now

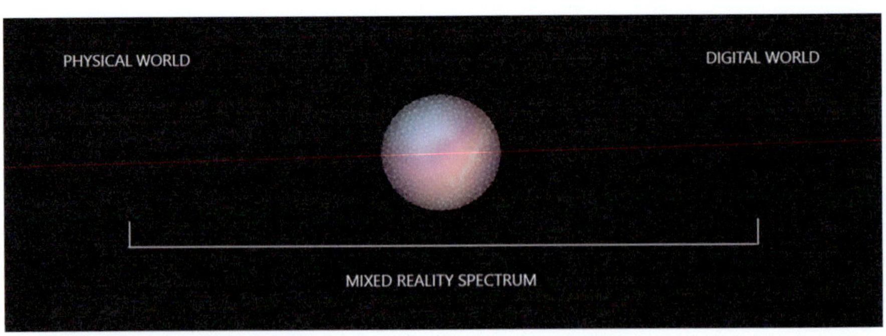

Fig. 12.5 Mixed reality is a blend of physical and digital worlds [10]

Fig. 12.6 A worker wearing an assisted reality device [12]

ubiquitous thanks to mobile devices and standalone smart speakers. Apple's Siri, Amazon's Alexa, Google's Assistant, and Microsoft's Cortana are all voice-driven software interfaces that are continuously gaining new capabilities. Many XR devices enable user control with handheld controllers, which have capabilities beyond button press inputs. Both voice-driven interfaces and human–computer interactions have been developed specifically for XR devices, including gaze and gesture controls [13]. Figure 12.7 compares conventional computing with extended reality [13].

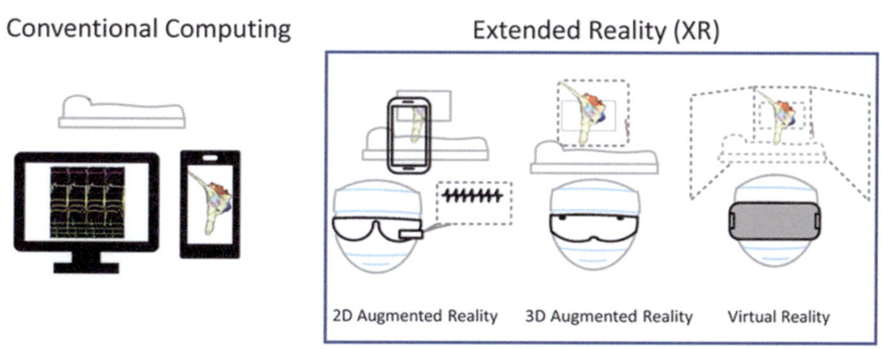

Fig. 12.7 Comparing conventional computing with extended reality [13]

12.3 Immersive Technologies in Architecture

In a world where technology continually blurs the boundaries between the real and the virtual, architecture is undergoing a revolutionary transformation. Architecture has gradually changed from traditional 2D drawings and physical models in recent years.

Immersive architecture is at the forefront of this gradual change, acting as a bridge between two distinct worlds: the physical environment we inhabit and the virtual landscapes we explore. It is redefining the concept of architecture design and revolves around creating environments that go beyond the static confines of traditional spaces. It weaves the physical and virtual realms together, immersing individuals in experiences that are both tangible and intangible.

Immersive architecture takes the concept of architectural design to a whole new level by creating environments that fully immerse the user in a sensory experience. Immersive technology in architectural design allows the creation of virtual project representations where space use can be optimized and potential design problems can be identified. Virtual reality (VR), augmented reality (AR), and mixed reality (MR) technologies play a significant role in immersive architecture, allowing architects to figure the exterior architecture design that transport occupants to other worlds or provide unique and captivating experiences. Architects can use VR for conducting user studies, presenting collaborative designs, building management, and design education. AR can be utilized for client communications in the same universal way that email, video calls, and 3D design software have in the past. Unlike VR, which creates a totally new and independent environment of the real world, AR includes virtual elements that interact with what already exists. MR combines the real-world environment and a virtual content world, enabling architects to collaborate in real time using holograms. Immersive technology tools and software often need other supporting technologies like headsets, 3D displays, 3D audio, gesture recognition readers, spatial sensing, speech recognition devices, haptics, drones, cameras, and omnidirectional treadmills that are used to fully connect with the various immersive technologies [14]. Figure 12.8 shows an example of how ImT is used in architecture [1].

As virtual, augmented, and mixed reality technologies continue to advance and be integrated into the architecture industry, there may be a learning curve involved for many in the industry. These immersive technologies in architecture have begun to be used in the design process.

Fig. 12.8 How immersive technology is used in architecture [1]

12.4 Applications of Immersive Technologies in Architecture

The potential applications and future directions for immersive architecture are vast and exciting. Immersive technologies have integrated many fields, including gaming, entertainment, education, training, healthcare, architecture, retail, real estate, and design. We can anticipate its further integration into urban planning, public spaces, design visualization, client engagement, and even personal homes. This could lead to the creation of smart cities with immersive urban experiences, sustainable buildings with adaptive interiors, and personalized virtual environments tailored to individual preferences. Common areas of applications include the following [15–17]:

- *Design Visualization*: Virtual reality offers architects an unprecedented opportunity to bridge the gap between imagination and reality. Real-time visualization is crucial, not only for speeding up workflows but also for maintaining the creative momentum that is essential in architectural design. Traditionally, architects relied on 2D drawings and static 3D models to convey their design concepts. VR, however, empowers architects to create immersive, interactive 3D environments that provide a true sense of scale, proportion, and spatial relationships. Through VR, architects can navigate through their designs as if they were physically present within them. It aids in making design decisions, as architects can experience the design from various angles, lighting conditions, and even simulate user interactions. A typical design visualization is shown in Fig. 12.9 [18].
- *Spatial Design*: As virtual worlds and spatial experiences evolve, our holistic perception of design is gradually changing. Design principles currently re-shaping Web3 and XR are also influencing real architectural environments. Apple has done

Fig. 12.9 A typical design visualization [18]

a great job in providing design standards for spatial user interfaces. An example of spatial design using ImT is shown in Fig. 12.10 [19].

Fig. 12.10 A typical spatial design using immersive technology [19]

- *Client Engagement*: One of the key advantages of incorporating virtual reality in architecture is the ability to actively engage clients in the design process. For clients, understanding architectural plans and visualizing the end product can be challenging, as they often lack the technical expertise to interpret traditional architectural drawings. VR solves this problem by offering a highly intuitive and immersive way for clients to explore and understand architectural designs. With VR, clients can put on a headset and walk through their future space, experiencing the design firsthand. They can provide feedback on the design in real-time, enabling architects to make adjustments and refinements on the spot.
- *BIM with AR*: The vast amount of information that a BIM model holds within it is difficult to structure and exchange, which can affect efficiency at a construction site in addition to a huge amount of time for site workers. However, AR is an effective way of retrieving the information and providing a seamless medium to project the information. The vast amount of data and interaction involved in the construction project can be accessed easily by AR, thus making it an important and seamless possible way for detecting aspects involved with the view of a user. The amalgamation of BIM with AR in particular and with ImTs in general is not only leveraging the smart and convenient way to be adopted in the AEC industry, rather they have a considerable impact on many other facets, namely the business promotion of projects, research studies in institutions, and enhancing the education of AEC professionals.
- *GAMMA Augmented Reality*: This is a building site monitoring application that uses augmented reality technology to overlay BIM 3D buildings via smartphones or tablets. It enables 3D BIM models to be viewed before and during the building process, creating an understanding of planning, avoiding errors, and reducing construction costs.
- *Simulation*: Immersive technologies enable architects and engineers to analyze potential issues before construction begins. Simulation tools using AR/VR help designers predict how a building will perform in the real world. By incorporating data such as weather patterns, traffic flows, crane and pedestrian movements, engineers can test the limits of a design before a brick is even laid. By using immersive virtual mock-ups, professionals can identify and address potential challenges more efficiently. Simulating real-world scenarios such as lighting conditions or energy performance can aid in the optimization of the building design. Developers can let stakeholders and end-users walk through new buildings without being physically on-site or immerse themselves in simulated environments.
- *Real Estate*: ImT has been a game-changer in various industries, and its applications in the real estate sector are no exception. In the real estate industry, immersive technology has made it increasingly possible to provide potential buyers with realistic, interactive experiences of properties. By utilizing VR and AR, real estate agents can showcase their properties to clients located anywhere in the world without the need for physical viewings. This not only saves time and money but also opens up new possibilities for international clientele, allowing the industry to expand its reach.

- *Home Tours*: Using VR/AR, prospective buyers can explore properties remotely. This convenience allows for a more efficient and streamlined viewing process, benefiting both the agent and the buyer. For example, AR can revolutionize the way potential buyers visualize properties by providing realistic, 3D representations of homes and their interior spaces.
- *Property Management*: AR tools can help property managers better understand their assets, monitor the state of their properties, and make informed decisions about maintenance and renovation requirements. ImT can enhance the communication and collaboration between property managers, landlords, tenants, and construction teams during the planning and execution of projects, ultimately improving efficiency and outcomes.
- *Education and Training*: Immersive technologies are making inroads into architectural education and professional training. They can be used in architecture education to help students experience spaces that they might not otherwise be able to. Architecture students can use VR to explore historic buildings, iconic structures, and renowned architectural designs from around the world, offering them valuable experiential learning opportunities. Immersive technologies offer new possibilities in education and training. They create immersive learning environments that simulate real-world scenarios for training purposes, enhance visualization and understanding of complex concepts, and provide interactive and hands-on experiences. ImT can be used to train real estate agents, architects, and other professionals in a more engaging and interactive way. This can help them better understand complex concepts and improve their skills.

12.5 Benefits

Virtual reality and augmented reality can improve design reviews, client presentations, spatial planning, virtual prototyping, and product design. With the advent of augmented reality (AR), architects will have new opportunities to blend the virtual and physical environments, offering enhanced design and construction experiences. Other benefits include the following [20]:

- *Collaboration*: In the architecture and construction industry, collaboration is a fundamental aspect of successful project delivery. VR has emerged as a transformative tool for enhancing collaboration among architects, designers, engineers, and other stakeholders. It overcomes the limitations of traditional communication methods by enabling real-time, interactive, and immersive collaboration. Architectural projects often involve a myriad of professionals with diverse expertise. With VR, these professionals can virtually come together in a shared digital environment, regardless of their physical locations.
- *Communication*: Communication with stakeholders in the AEC industry has always been heavily reliant on visual means such as sketches, two dimensional (2D) drawings and images. Further advancements in technology, such as building information modelling (BIM) and immersive technologies, made it possible to

project designs in three-dimensions (3D), which profoundly revolutionized the industry.

- *Cost Reduction*: Immersive technologies, such as VR, enable architects and real estate professionals to create realistic simulations of properties and spaces, through virtual property tours, virtual staging, design visualization, collaborative design, training and education, reducing the need for physical models and on-site visits. This leads to significant cost savings in the long run.
- *Efficiency*: Efficiency is a critical factor in architecture, where time and resources are often limited. Immersive technology enables faster decision-making as virtual reality enables stakeholders to make quicker and more informed decisions about design elements and property investments. Virtual reality plays a vital role in streamlining the design process, reducing errors, and saving both time and money. Through VR, architects can conduct design iterations rapidly and assess the impact of design changes immediately. VR aids in minimizing construction errors.
- *Sustainability*: Immersive technologies like VR, AR, and BIM are revolutionizing sustainable architecture by enabling architects to incorporate eco-friendly principles from the very beginning. VR can be instrumental in sustainability considerations. Architects can use VR to simulate various environmental conditions, such as natural lighting and thermal comfort, to optimize energy efficiency and reduce the building's ecological footprint.
- *Storytelling*: Immersive architecture is not only about aesthetics but also about storytelling and creating emotional connections between the architecture and its occupants. As a storytelling medium for architects, VR allows individuals to immerse themselves in the atmosphere, explore, and interact with the functionality and purpose of a digital representation of a proposed design. It encourages self-expression, customization, and the exploration of individual needs.
- *Infrastructure Design*: Immersive technologies help government planners visualize and assess proposed infrastructure projects, such as a new bridge, allowing stakeholders to explore 3D models to simulate scenarios and understand the potential impact on the environment and communities before implementation.
- *Heritage Preservation*: Immersive technologies can preserve and showcase cultural heritage sites and artifacts by allowing users to virtually explore historical sites, museums and archaeological sites, providing access to cultural experiences and artifacts that may be physically inaccessible or at risk.
- *Marketing*: The potential for VR in architectural marketing and presentations is substantial. Real estate developers, for example, can use VR to showcase properties to potential buyers, providing immersive experiences that convey the atmosphere and functionality of spaces that are yet to be built.
- *Smart Materials*: The intersection of light, material, and architecture is at the heart of frameless immersive art experiences. The development of smart materials (materials that can change properties in response to external stimuli) offers new possibilities for ImT. These materials can be designed to change color, transparency, or texture in response to light, temperature, or touch, creating dynamic, responsive environments.

12.6 Challenges

Modernization was one of the biggest challenges faced by the AEC industry. Recently, however, the industry has been exposed to immersive technologies. Immersive technologies come with challenges like cost and technical expertise. With the expected increase of integration of immersive tools as the technology advances, it will become more accessible, with lower costs, better hardware, and more user-friendly software. To develop VR content that is truly engaging and compelling is a challenge that requires a considerable set of skills as well as resources. Other challenges include the following [20]:

- *Cost*: The cost is one of the major challenges that could restrain accessibility to ImT devices. The high cost of peripherals is a major concern for the wide adoption of ImT across various industries. Immersive technologies are expensive to implement since the cost of both procurement and maintenance of sophisticated devices and applications become prohibitive. The cost of immersive technologies may be too high to convince management to adopt it. Newer technologies rapidly make older systems obsolete and increase their life-cycle costs. As technology evolves, VR hardware becomes more accessible and affordable, making it a viable tool for architects of all scales and budgets.
- *Side Effects*: The use of immersive technologies can bring side effects to the user. The most common being nausea, similar to motion sickness, as the brain cannot correlate the motion observed by the eyes with that of the body standing still in a room. Physical discomfort and ocular fatigue may also occur from prolonged use of immersive devices.
- *Ethics*: Ethical considerations around privacy and data security will become increasingly important as VR is used in sensitive architectural applications. Ethical issues also include prolonged exposure to the virtual environment resulting in users facing difficulties in performing normal tasks in the real world.
- *Emerging Technologies*: Designers can leverage emergent technologies to create and design new structures. For example, virtual reality (VR) and artificial intelligence (AI) are rapidly transforming the way we design and experience architecture. VR enables remote collaboration, while AI is a game changer for the residential architecture, real estate, and design industry.
- *Error*: As the AEC industry is accident-prone, the incorrect interpretation of information is common, mainly due to a less skilled workforce, which eventually increases the cost of a project, reduces the quality, and delays the schedule.

12.7 Conclusion

Immersive technology is already impacting architectural practice in all phases of the project lifecycle, from its use as a visualization platform to its function as an analytical tool. The significance of immersive technologies in architecture continues to grow, and the future holds even more promise. As technological advances expand, these immersive experiences reshape the ever-growing architectural field. Architects who embrace these technologies will be better positioned to create innovative, efficient, and sustainable structures that meet the ever-evolving needs of society.

Both interactive and immersive architecture concepts are instrumental in shaping the future of architecture design. They enable modern architects to push the boundaries of creativity, offering endless possibilities for creating spaces that are not only visually striking but also highly functional and engaging. More information on the integration of immersive technology into the architecture industry is available from the books in [21–26] and the following related journals:

- *Technologies*
- *Automation in Construction*

References

1. R. Nieminen, Immersive technology may be architecture's best tool for communication (2018). https://www.archpaper.com/2018/05/immersive-technology-architectures-best-tool-communication/
2. M.N.O. Sadiku, P.A. Adekunte, J.O. Sadiku, Immersive technologies in architecture. Int. J. Trend Sci. Res. Dev. **9**(1), 367–376 (2025)
3. M.N.O. Sadiku, C.M.M. Kotteti, S.M. Musa, Augmented reality: a primer. Int. J. Trend Res. Dev. **7**(3) (2020)
4. What is augmented reality or AR? https://dynamics.microsoft.com/en-us/mixed-reality/guides/what-is-augmented-reality-ar/
5. L. van Heerden, What is extended reality? (2021). https://journeyapps.com/blog/what-is-extended-reality/
6. A. Xperteye, What is assisted reality? Here is what you need to know (2022). https://blog.amaxperteye.com/what-is-assisted-reality-here-is-what-you-need-to-know
7. M.O. Onyesolu, F.U. Eze, Understanding virtual reality technology: advances and applications. Adv. Comput. Sci. Eng. 53–70 (2011)
8. Immersive technologies explained. https://www.csps-efpc.gc.ca/tools/articles/immersive-technologies-eng.aspx
9. M. Singh, M.P. Singh, Augmented reality interfaces. IEEE Internet Comput. 66–70 (2013)
10. What is mixed reality? (2023). https://learn.microsoft.com/en-us/windows/mixed-reality/discover/mixed-reality
11. Mixed reality, *Wikipedia*, the free encyclopedia. https://en.wikipedia.org/wiki/Mixed_reality
12. What is assisted reality? Here is what you need to know (2022). https://blog.amaxperteye.com/what-is-assisted-reality-here-is-what-you-need-to-know
13. C. Andrews et al., Extended reality in medical practice.Curr. Treat Options Cardiovasc Med. **21**(4) (2019)
14. Architecture trends: Immersive technologies. https://www.re-thinkingthefuture.com/architectural-styles/a11643-architecture-trends-immersive-technologies/#:~:text=Immersive%20technologies%20provide%20an%20improved,relatively%20uncommon%20in%20architectural%20practice.
15. The significance of virtual reality in architecture: An immersive evolution. https://www.cuubstudio.com/blog/the-significance-of-virtual-reality-in-architecture-an-immersive-evolution/#:~:text=Virtual%20Reality%20(VR)%20has%20emerged,and%20enhance%20overall%20project%20efficiency.

16. A. Prabhakaran, A. Mahamadu, L. Mahdjoubi, Understanding the challenges of immersive technology use in the architecture and construction industry: a systematic review. Autom. Constr. **137** (2022)
17. Immersive technology in architecture. https://drawandcode.com/sectors/immersive-in-architecture/
18. M. Terenzio, Immersive spatial design- Unifying the best of the physical and virtual world. https://interiorarchitects.com/immersive-spatial-design/
19. G. Boulet, Immersive technologies explained (2023). https://www.csps-efpc.gc.ca/tools/articles/immersive-technologies-eng.aspx
20. Immersive architecture: A new connection between two worlds of reality and virtual. https://huurstudios.com/immersive-architecture-connection-reality-virtual/
21. A. Prabhakaran et al. (eds.), *Applications of Immersive Technology in Architecture, Engineering and Construction: A Handbook (Spon Research)* (Routledge, 2025)
22. L. Psarologaki, *Site-Reliant Immersive Experiences: Sensing Affective Spaces in Art and Architecture* (Bloomsbury Visual Arts, 2025)
23. S. Eloy, A. Kreutzberg, I. Symeonidou (eds.), *Virtual Aesthetics in Architecture: Designing in Mixed Realities* (Routledge, 2021)
24. A. Kreutzberg, I. Symeonidou, S. Eloy (eds.), *Virtual Aesthetics in Architecture: Designing in Mixed Realities.* (Taylor & Francis, 2021)
25. M.A. Schnabel, X. Wang (eds.), *Mixed Reality In Architecture, Design* (Springer, And Construction. Netherlands, 2009)
26. D. Bertol, D. Foell, *Designing Digital Space: An Architect's Guide to Virtual Reality* (Wiley, 1997)

Index

M. N. O. Sadiku, *Emerging Technologies in Architecture*,
https://doi.org/10.1007/978-3-031-99509-5

MIX
Papier aus verantwortungsvollen Quellen
Paper from responsible sources
FSC® C105338

If you have any concerns about our products,
you can contact us on
ProductSafety@springernature.com

In case Publisher is established outside the EU,
the EU authorized representative is:
Springer Nature Customer Service Center GmbH
Europaplatz 3, 69115 Heidelberg, Germany

Printed by Libri Plureos GmbH
in Hamburg, Germany